Amos Arthur Heller

Botanical Explorations in Southern Texas During the Season

of 1894

Amos Arthur Heller

Botanical Explorations in Southern Texas During the Season of 1894

ISBN/EAN: 9783744717052

Printed in Europe, USA, Canada, Australia, Japan

Cover: Foto ©Andreas Hilbeck / pixelio.de

More available books at **www.hansebooks.com**

CONTRIBUTIONS

FROM THE

...of Franklin and Marshall College,

LANCASTER, PA.

No. 1.

...L EXPLORATIONS

IN

...RN TEX...

PRIC...

ISSUED FEBRUARY 6, 1895.

THE NEW ERA PRINTING HOUSE,
1895.

CONTRIB[UTIONS]

FROM THE

Herbarium of Franklin a[nd Marshall College]

LANCASTER[, PA.]

No. 1.

BOTANICAL EXPL[ORATIONS]

IN

SOUTHERN [CALIFORNIA]

DURING

THE SEASON [OF 1894]

By A. A. HEL[LER]

PRICE, $1.00.

ISSUED FEBRUARY 6, 1895.

THE NEW ERA PRINTING HOUSE,
1895.

INTRODUCTION.

The State of Texas, an empire in itself, comprising one-twelfth of the area of the United States, with great diversity of soil and climate, must necessarily present a corresponding diversity in plant life. When all of its immense area is thoroughly explored, it will undoubtedly yield as many or perhaps more species than are contained within the limits of Gray's Manual.

As is stated in the Introduction to the Botany of the Mexican Boundary Survey, a line drawn from the Pan Handle southeast to the vicinity of Corpus Christi, divides the State into two unequal portions. The smaller eastern part is well watered as a whole, and gives rise to more luxuriant vegetation than the other, where rain is uncertain and often scanty. The flora of this eastern section, at least in its northern and eastern portions, has many species in common with the adjacent States of Louisiana and Arkansas and the Indian Territory.

The larger southern and western division, in many places a veritable desert, contains many peculiar plants, found only within its limits, especially in the central portions, in the limestone foothills, and in the mountains of the extreme western part. Along the southern border, the species are essentially Mexican, intermingling in the mountain region with more northern forms found in the mountains of New Mexico, and with characteristic Texan plants.

Undoubtedly the best part of the State botanically, and also the least known, is the extreme southern portion, from Corpus Christi west to Laredo and south to Brownsville. In the Mexican Boundary Survey, the statement is made that "the botany of this region is too well known from various published accounts to require further details in this place."

Nevertheless, the fact remains that practically we know nothing of what it contains. The bulk of Neally's new species were collected between Brazos Santiago and Rio Grande City. Between the latter place and Laredo I find no mention of plants ever having been collected. The rough character of the country, its inaccessibility, dryness, and the great expense to be incurred in properly working it up, are formidable barriers to an individual explorer.

Corpus Christi Bay and the lower Nueces may be considered the northern boundary of this interesting tract of country. The vegetation so characteristic of a dry country is very marked, nearly everything being armed with thorns or spines. At Corpus Christi, where in all, eight weeks were spent, one can form a very good idea of what the country between there and the Rio Grande is like. The lower, or business portion of the town, is built where once the waters of the bay rolled. This low portion ends rather abruptly on the south side, where the bluff rises precipitately to a height of thirty or forty feet. It gradually becomes broader, until its northern portion is about two miles wide. The bulk of this width, though, is an arm which juts out, forming the division between Corpus Christi Bay and Nueces Bay. Its extreme northern portion is low and flat, marshy near the water, and at times of unusually high water, overflowed. The plants which occur on these "Flats" are those which are usually found in saline soil, such as *Batis maritima*, *Suaeda suffrutescens*, etc. Here the ascent to the plateau is less abrupt, and the plateau itself lower. Much of this low land, at least near the beach, is composed of a shell deposit, instead of sand. One can dig into it for a distance of several feet without meeting with anything else.

Corpus Christi Bay is very shallow, the maximum depth being seventeen feet, but the average is hardly ten. Nueces Bay is still shallower and full of quicksand. I am told that long ago Spanish ships of the largest size were accustomed to cast anchor where now a boat with a draught of three or four feet would run aground. At present the only deep water inlet is Aransas Pass, at the upper end of Mustang Island. It seems, therefore, that a gradual elevation of the land is slowly going on.

The plateau is level, cut here and there near the shore by arroyos, or ravines. Near these the ground is full of holes, which gradually become larger, until finally they cave in, forming side branches. The soil is very rich, and is called "black land," or "black waxy land." When wet by rain it becomes exceedingly sticky, great clods of it clinging to one's shoes, so that walking becomes extremely tiresome. It is covered with sod, and under favorable conditions would be a splendid agricultural country; but lack of rain makes the raising of crops uncertain, and during the past year many cattle perished from starvation.

Until last April, there had not been a good rain for over four years, nothing but light showers at intervals, and these scarcely sufficed to moisten the ground. Early in April there were two or three heavy rainstorms, but storms of this kind occurring at long intervals do very little good and often much damage.

During the spring months there is a great deal of cloudy weather, and in the coast region, one would expect to have an abundance of rain. The strong trade winds which blow almost continually at that time probably carry the clouds away before they have had time to deposit their moisture.

The features so characteristic of the Atlantic coast region are here entirely wanting. Swamps are conspicuous by their absence, as likewise are trees, which are found only along the upper end of Nueces Bay. Occasionally *Prosopis juliflora*, the Mesquite, becomes large enough to be called a tree, but even then is low and spreading. Indeed, low and spreading, stunted-looking trees are the rule, as the tall graceful forms of a more northern climate do not find a place here; but bushes of various kinds are abundant, forming the dense and usually impenetrable chapparral. This chapparral is a very deluding thing, too. One ventures into it by way of one of its lanes, which here and there sends off side branches, imagining that by one of them he will find an exit, only at last to discover that the way is completely blocked by a solid mass of bushes. Getting lost in a place of this sort would be a very serious matter.

Many species of smaller plants are found only under the chapparral. They have either betaken themselves to these places of safety for self-protection, or are the remnant of a flora which once thickly dotted the open places.

When grass is scarce the cattle become omniverous, devouring anything that they can chew, whether it be good, bad or indifferent; but into the thorny wilderness they cannot penetrate. The collector, in order to be successful in obtaining good specimens from these places, should possess a great amount of patience, go prepared to cut down the bushes, and if he is inclined to profanity will probably exhaust his vocabulary before finishing the job.

Nine miles southeast of Corpus Christi is the Oso, a salt water lake, three or four miles in diameter, connected with the bay by two inlets about fifteen feet wide, and distant from each other about a mile. On all maps which I have seen, the Oso is marked as an arm of the bay instead of being separated from it by a strip of land, which at the "Blind Oso," its narrowest part, is about 150 yards wide at times of unusually high water. Beyond this narrow strip of dry sand, is a mud flat, a half mile wide, before the waters of the Oso are reached. At other places the strip of land separating the two bodies of water is much wider, being almost a mile on the south side of the Blind Oso.

Between the Oso and the Lagoon de Madre is a strip of slightly elevated sandy land, about three miles wide, called Flower Bluff, the principal growth of which is the live oak, reduced to a scrubby bush, from three to eight feet high. Several truck farms are located on it, and although there is more moisture here than at Corpus Christi the vegetables produced are often of an inferior quality.

In all, eight weeks were spent at Corpus Christi, from March 3d to

April 17th, and from May 28th to June 9th. The part explored was a narrow strip along the shore, from the mouth of the Nueces River to Flower Bluff, a distance of twenty-five miles. One day was spent in San Patricio county, across the bay from Corpus Christi. Away from the immediate vicinity of the bay the country is too inaccessible on account of the chapparral, and does not contain enough moisture to make good collecting ground, although it might if there were not several head of cattle to every plant which ventures above ground.

The bulk of the collecting was done in and about the town. Two trips were made to the mouth of the Nueces along Nueces Bay, two to the Oso, and one to Flower Bluff.

When I arrived there early in March, plants were plentiful and blooming profusely after the slight winter rains. During the last week of March a "norther" came down, followed by another in a few days, when, as if by magic, the plants began to droop, the flowers to disappear, and on some of the pasture land scarcely a sprig of green could be seen—nothing but the brown, bare earth.

The most prominent herbaceous plants on the plateau in early spring were *Lesquerella Gordoni* and a species of *Œnothera* as lately received, apparently close to *Œ. primiveris*. Of shrubs, the most common were *Prosopis juliflora, Castela Nicholsoni, Celtis pallida, Zizyphus obtusifolia, Acacia amentacea, A. tortuosa* and *Colubrina Texensis*.

On April 17th I moved to Kerrville, 71 miles northwest of San Antonio, and about 280 from Corpus Christi. It is a small town of perhaps 1000 inhabitants, situated on the headwaters of the Guadalupe, at an elevation of 1650 feet above sea level, and is one of the health resorts of Texas. It is situated in a limestone formation and surrounded by hills, the highest of which are about 2000 feet above sea level. Occasionally one of these hills is isolated and cone-shaped, like the buttes of the Bad Lands in the Dakotas. They are likewise terraced, a wall of rock two or three feet high completely encircling the hill; above this a bench of earth, then another rock wall, and so on to the summit, the intervals between the benches becoming less as one ascends.

At the northwestern end of the town, the Guadalupe receives a small tributary called Town Creek. My explorations here were confined to the immediate vicinity of Kerrville, along the banks of the Guadalupe for a distance of about two miles, along the banks of Town Creek for about a mile, and the surrounding hills, principally those on the left bank of the river, and at a distance of about a mile from the town. One day was spent along Bear Creek, in the extreme northeastern part of the county, one trip made to a point on Wolf Creek, about fourteen miles north of

Kerrville, and one trip up Town Creek for a distance of about seven miles.

For at least twenty-five miles on all sides, and for many more in some directions, the same limestone formation prevails, and plant life appears to be pretty uniform throughout. Having found by experience that long trips yielded practically the same things which I could find within a radius of a mile from Kerrville, I directed my attention to thoroughly exploring a small area, dividing the time so that each particular place was visited once a week.

The steep, stony, left bank of the river for a distance of about one-fourth of a mile took up at least two days of the week, Town Creek one or two, and the hills the balance of time. To the best of my knowledge, horehound, which is abundant about the streets, and a species of Juniper occasionally met with along Town Creek, are the only plants found in flower or fruit between the middle of April and the first of July that I did not collect.

The characteristic plants of the limestone region in Kerr county are many. In fact the bulk of the species are plentiful over the whole area at certain elevations. At no place is there a greater range of more than 400 feet between the lowest and the highest elevations, yet a number of species growing abundantly on all the hilltops are not found at the lowest elevations, and some of them only on the summits. *Acacia Roemeriana, Coreopsis Drummondii Thelesperma filifolium, Bifora Americana*, and *Brazoria scutellarioides* are examples of the hilltop flora. A few of those found at both the highest and lowest elevations are *Lesquerella recurvata, Kuhnistera pulcherrima, Cassia Roemeriana* and *Erigonum longifolium*. Along streams, the dwarf walnut, *Juglans rupestris*, is very plentiful.

Three trips were made to San Antonio, with very satisfactory results, the rich, moist ground along the river banks always producing an abundance of plants.

In all, 573 numbers were collected, only a few of them being duplicated. Of these, 39 were collected at San Antonio, 4 at Kenedy, Carnes county, 8 at Waco, McLennan county, 248 about Corpus Christi, and 299 about Kerrville.

The orders represented by the largest number of species are Compositae and grasses, of which some 60 species each were collected. Texas is very prolific in grasses, but they grow in scattered clumps or as solitary plants, rarely forming a sward, as they do further north. The Leguminosae are represented by over 50 species, many of which are very plentiful. Next come Euphorbiaceae with about 25 species, Labiatae with 20, and Umbelliferae and Cruciferae with 12 or 15.

As before mentioned, Corpus Christi Bay and the lower Nueces may be considered the northern boundary of a flora peculiar to southern Texas. Within the limits of this, six of my eleven new species were collected, and a number of rare ones re-discovered.

Between this and the more elevated limestone district, is a tract 200 miles wide at some places, which appears to have a flora more or less distinct from the other two sections, although many species common to both are found within its limits. My knowledge of this central tract was obtained principally from observations made along the railroad while traveling. Here the Mesquite is much more abundant than in the southern section, while in the limestone region proper it seems to be entirely wanting.

At San Antonio, or rather a few miles west, the limestone formation and hilly ground begins, with a distinct and sharply marked flora, the Mesquite and other plants so common and rather monotonous suddenly giving place to groups of *Sapindus, Monarda citriodora*, etc.

My thanks are due to Mr. Frederick V. Coville, Government Botanist, who kindly gave me the opportunity of making my determinations at the U. S. National Herbarium, and determined the *Junci;* to Mr. J. N. Rose, who determined the Umbelliferae; to Prof. F. Lamson Scribner, for the determination of the grasses; to Dr. N. L. Britton, for the use of the botanical library of Columbia College and of the herbarium while making corrections and final determinations; to Mr. John K. Small, for the determination of the Polygonums and Rumex; to Dr. Thos. C. Porter, for the determination of 100 of the first numbers, and to Mr. M. S. Bebb, for the determination of the willows. Without the aid of these gentlemen, it would have been impossible to produce the present work.

In the matter of citations of publication, it is to be regretted that in perhaps a dozen cases the date could not be obtained. Certain numbers of the *Botanical Magazine* and *Botanical Register* could not be consulted, as well as several other works.

Apology must be made for the unnecessary omission of many type localities. The first part of the enumeration was prepared before I had access to the original descriptions, and later this omission was overlooked.

The last line in the paragraphs under the enumeration and descriptions of species refers to the date of collection and the number of the plant, the latter in parenthesis, as "May 26 (1781)."

Enumeration and Descriptions

—OF—

SPECIES.

FILICES.

ADIANTUM L. Sp. Pl. 1094 (1753).
Adiantum Capillus-Veneris L. Sp. Pl. 1096 (1753).
On a dripping limestone bluff on the left bank of the Guadalupe, at Kerrville, where it was plentiful for a distance of about 400 yards. Altitude 1625 feet. The only favorable situation in that region for its growth.
July 2 (1939).

DRYOPTERIS Adans. Fam. Pl. 2: 20 (1763).
Dryopteris patens (Swartz) Kuntze, Rev. Gen. Pl. 813 (1891).
Aspidium patens Swartz, in Schrader's Jour. 2: 26 (1800).
On the banks of the San Antonio just below the S. P. bridge, altitude 600 feet. Only a few plants were seen.
June 9, (1835); type locality, W. Jamaica.

CONIFERAE.

TAXODIUM (L.) L. C. Rich. Ann. Mus. Paris 16; 298 (1810).
Taxodium distichum (L.) L. C. Rich. Ann. Mus. Paris 16: 298 (1810).
Cupressus disticha L. Sp. Pl. 1003 (1753).
A number of fine large trees on the upper Guadalupe, at Kerrville, altitude 1600 feet. No "knees" were observed, and usually the trees were found growing on the dry banks. This, and a species of Juniper unfortunately not collected, were the only Conifers in the region.
April 19 and June 20 (1620).

GRAMINEAE.

ANDROPOGON L. Sp. Pl. 1045 (1753).
[SORGHUM Pers. Syn. 1: 101 (1805).]
[CHRYSOPOGON Trin. Fund. Agrost. 187 (1820).]

Andropogon Halapensis (L.) Brot. Flor. Lusit. 1: 89 (1804).
Holcus Halapensis L. Sp. Pl. 1047 (1753).
Sorghum Halapense Pers. Syn. 1: 101 (1805).

Common in cultivation under the name of "Johnson grass," but naturalized in many places. Vigorous plants were seen on the stony banks of the Guadalupe, altitude 1630 feet, at a distance from cultivated ground. San Antonio, Bexar county, altitude 600 feet, on the edge of a cultivated field.

May 5 (1706).

Andropogen saccharoides Sw. Fl. Ind. Occ. 1: 205 (1797).

A few plants collected in a grassy level place along the San Antonio, altitude 600 feet, but an abundance of it was seen on the stony sloping left bank of the Gaudalupe, at Kerrville, altitude 1630 feet.

May 5 (1704); type locality, S. Jamaica.

NAZIA Adans. Fam. Pl. 2: 31 (1763).
[TRAGUS Hall. Hist. Stirp. Helv. 2: 203 (1768).]
[LAPPAGO Schreb. Gen. 55 (1789).]

Nazia racemosa (L.) Kuntze, Rev. Gen. Pl. 780 (1891).
Cenchrus racemosus L. Sp. Pl. 1049 (1753).
Lappago racemosa Willd. Sp. Pl. 1: 484 (1798).

Growing prostrate in the sand, along the beach of Corpus Christi Bay, the spreading plants growing comparatively close together. Seen at only one place near the upper end of the Bay at sea level.

May 29 (1794).

PASPALUM L. Syst. Ed. 10, 2: 855 (1759).

Paspalum pubiflorum Rupr. ex Galeotti, Bull. Acad. Brux. 9: 237 (1842).

At San Antonio in cultivated ground it was rather stout and inclined to be prostrate, while in rich shady ground along Town Creek at Kerrville it grew in clumps with long, spreading ascending stems, two or three feet long.

San Antonio, Bexar county, May 5 (1699); Kerrville, Kerr county, June 16 (1872).

Paspalum setaceum Michx. Fl. Bor. Am. 1: 43 (1803).
Paspalum pubescens Muhl. Gram. 92 (1817).
Paspalum ciliatifolium Michx. Fl. Bor. Am. 1: 44 (1803).

A few plants were found along the railroad at Corpus Christi, and at Flower Bluff in sand, at sea level. This is the *P. ciliatifolium.*
April 9 (1546).

PANICUM L. Sp. Pl. 55 (1753).

Panicum autumnale Bosc.; Spreng. Syst. 1: 320 (1825).
Panicum fragile Kunth, Rev. Gram. 1: 36 (1829).?
Panicum divergens Muhl. Gram. 120 (1817)? teste Chapman.

Rather common in stony limestone ground about Kerrville, especially along the river, altitude 1630-1700 feet.
May 14 (1744).

Panicum colonum L. Syst. Ed. 10, 105 (1784).

Prostrate, growing in depressions at Corpus Christi, which in wet weather are filled with water, altitude 40 feet. Seen at only one place, and not plentiful. March 26 (1501). Also along the left bank of the Guadalupe at Kerrville, altitude 1600 feet, on flat rocks covered with a thin layer of earth. At times of high water these rocks are evidently submerged. Prostrate, rosette-like in habit, the culms often nearly two feet long. Leaves usually marked laterally with purplish bands.
June 27 (1923).

Panicum dichotomum L. Sp. Pl. 58 (1753).?

Growing in dense tufts in gravel along the left bank of the Guadalupe at Kerrville.
June 19 (1888).

Panicum fuscum Swartz, Prodr. Veg. Ind. Occ. 23 (1783-87).

A handsome yellowish-green grass, erect, growing in slender tufts. Noticed only in cultivated ground at San Antonio, altitude 600 feet, and on the edges of fields at Kerrville, altitude 1650 feet.
May 5 (1698)'; type locality, Jamaica.

Panicum maximum, Jacq. Ic. Pl. Rar 1: *t. 13* (1811-13).

A dense clump with culms about four feet long, growing on the edge of a field along the left bank of the Guadalupe, altitude 1625 feet. Within the range of Coulter's Manual of Western Texas, but not recorded in that work.
June 21 (1898); type locality, W. Indies.

Panicum nitidum Lam. Encycl. 4: 748 (1797).

Moist places in limestone ground along the left bank of the Guadalupe, at Kerrville, altitude 1600 feet. Seen at only one station. Usually solitary in growth and scattered. Recorded as occurring from "eastern Texas and eastward to the Atlantic."

May 16 (1752).

Panicum obtusum H.B.K. Nov. Gen. 1 : 98 (1815).

Abundant in limestone at Kerrville, altitude 1625-1650 feet; growing in cultivated fields, waste places about the streets and in yards.

May 14 (1741); type locality, Mexico, near Guanaxuato.

Panicum pedicellatum Vasey, Bull. No. 8, U. S. Dept. Agric. Div. Bot. 28 (1889).

Common on stony wooded hillsides about Kerrville, altitude 1625-1800 feet, growing in scattered tufts. The most northern station observed by myself was along the banks of Wolf Creek, fifteen miles north of Kerrville (No. 1726).

May 15 (1636, 1736, 1766).

Panicum Hallii Vasey, Bull. Torr. Club, 11 : 61 (1884).

Growing in gutters at Kerrville, and on flat exposed rocks on the Guadalupe, just below the town; altitude 1600-1650 feet. At Corpus Christi it was found in a sandy, open field. This plant was distributed as *P. proliferum*.

Corpus Christi, March 23 (1490); Kerrville, Kerr county, June 18 (1883).

Panicum Reverchoni Vasey, Bull. No. 8, U. S. Dept. Agric. Div. of of Bot. 25 (1889).

Very little seen, and apparently not plentiful in Kerr county. Growing on a bank along the roadside, on Town Creek, altitude 1625 feet.

April 19 (1603).

Panicum sanguinale L. Sp. Pl. 57 (1753).

Digitaria sanguinalis Scop. Fl. Carn. Ed. 2, 1 : 52 (1772).

Syntherisma praecox Walt. Fl. Car. 76 (1788).

Paspalum sanguinale Lam. Tabl. Encycl. 1 : 176 (1791).

About the streets of Kerrville, and growing very luxuriantly in wet places on the Guadalupe, where it occurs in tangled mats, the culms weak and reclining; altitude, 1600-1650 feet. Altogether much stouter and coarser than the plant found in waste places in the North.

June 27 (1917).

Panicum scoparium Lam. Encycl. 4 : 744 (1797). ?

Panicum pauciflorum Ell. Bot. S. C. & Ga. 1 : 120 (1817).

Growing in rich, shaded ground along Town Creek, at Kerrville, altitude 1600 feet.

May 19 (1759).

Panicum filipes Scribn. n. sp.

(PLATE I.)

Culms slender, about 30 inches high, branched at the base, minutely bearded at the nodes, otherwise smooth; sheaths smooth; ligule reduced to a minute barbate ring, or obsolete; leaf-blade smooth, the upper one on the flowering culm three to four lines wide, and six to eight inches long, those of the sterile shoots one to two lines wide, and five inches to nearly a foot long; panicle twelve inches long, the few-flowered capillary branches and long pedicels spreading; spikelets about 1¼ lines long, ovate acute; first glume broadly ovate obtuse, three-nerved, about one-half the length of the second; second and third glumes nearly equal, acute; the second five to seven-nerved, the third seven to nine-nerved, and with a palea about half its length; flowering glume nearly elliptical, obtuse, very smooth and shining, about one-fourth shorter than the second and third glumes. A slender, erect or ascending, smooth perennial (?), with long, narrow leaves, and ample capillary panicles. Description drawn from single specimen.

Growing in rich shaded ground in the upper part of the "Arroyo," at Corpus Christi, altitude 30 feet. Scarce.

May 31 (1809).

CHAMAERAPHIS R. Br. Prodr. Fl. Nov. Holl. 1: 193 (1810).

[SETARIA Beauv. Agrost. 113 (1812), not Ach. 1798.]

Chamaeraphis setosa (Swartz) Kuntze, Rev. Gen. Pl. 768 (1891).

Panicum setosum Swartz, Prodr. 22 (1783–87).

Setaria setosa Beauv. Agrost. 51 (1812).

In a cultivated field along the left bank of the Guadalupe, altitude 1625 feet. Called "wild millet" by the farmers. Not observed at any other place.

June 21 (1897).

Chamaeraphis glauca (L.) Kuntze, Rev. Gen. Pl. 767 (1891).

Panicum glaucum L. Sp. Pl. 56 (1753).

Setaria glauca Beauv. Agrost. 51 (1812).

A stout form with perennial root, found sparingly in gravelly places along the Guadalupe at Kerrville, altitude 1600 feet. Solitary, and scattered at long intervals.

June 19 (1889).

Chamaeraphis caudata (Lam.) Kuntze, Rev. Gen. Pl. 769 (1891).

Panicum caudatum Lam. Tabl. Encycl. 1: 171 (1791).

Setaria caudata R. & S. Syst. Veg. 2 : 495 (1817).

Found sparingly on a grass covered knoll, at the "Blind" Oso, about 100 yards from the beach.

March 21 (1480).

CENCHRUS L. Sp. Pl. 1049 (1753).

Cenchrus tribuloides L. Sp. Pl. 1050 (1753).

Cenchrus Carolinianus Walt. Fl. Car. 79 (1788), teste Pursh.

In dry, open sandy ground on the plateau at Corpus Christi, where it was plentiful. A slender prostrate form very different from our northern plant.

March 23 (1392).

ZIZANIOPSIS Doell et Aschers. in Mart. Fl. Bras. 2 ; Part 2, 12 (1871).

Zizaniopsis miliacea (Michx.) Doell et Aschers.; Baill. Hist. Pl. 12 : 293 (1893).

Zizania miliacea Michx. Fl. Bor. Am. 1 : 74 (1803).

Growing on the edge of the San Antonio in shallow water, at San Antonio, altitude 600 feet. Two clumps of it observed.

May 5 (1710).

PHALARIS L. Sp. Pl. 54 (1753).

Phalaris Caroliniana Walt. Fl. Car. 74 (1788), fide Munro.

Phalaris intermedia Bosc.; Poir. in Lam. Encycl. Suppl. 1 : 300 (1810).

Phalaris microstachya DC. Cat. Hort. Monsp. 131 (1813).

Phalaris Americana Ell. Bot. S. C. and Ga. 1 : 101 (1817).

Growing along the railroad tracks, near Gregory, San Patricio county, altitude 35 feet, and at San Antonio, altitude 600 feet.

April 17 (1578); type locality, Carolina.

ARISTIDA L. Sp. Pl. 82 (1753).

Aristida purpurea Nutt. Trans. Am Phil. Soc. (II.) 5 : 145 (1837).

Growing about ant hills in open sandy pasture land, near Gregory, San Patricio county, altitude 35 feet. Not seen except where ants had disturbed the ground.

April 14 (1579).

LIMNODEA Dewey, Cont. U. S. Nat. Herb. 2 : 518 (1894).

Limnodea Arkansana (Nutt.) Dewey, Cont. U. S. Nat. Herb. 2 : 518 (1894).

Greenia Arkansana Nutt. Trans. Am. Phil. Soc.(II.) 5 : 142 (1837), not W. & Arn.
Thurberia Arkansana Benth. in Vasey, Descr. Cat. Gr. U. S. 22 (1885).
Plentiful along railroad tracks near Gregory, San Patricio county, altitude 35 feet.
April 14 (1577).

AGROSTIS L. Sp. Pl. 6 (1753).

[TRICODIUM Michx. Fl. Bor. Am. 1 : 41 (1803).]
Agrostis verticillata Vill. Prosp. 16 (1779).
In wet limestone soil, left bank of the Guadalupe, at Kerrville, the stalks matted together; altitude 1600 feet.
May 14 (1742); type locality, Europe.

TRISETUM Pers. Syn. Pl. 1 : 97 (1805).

Trisetum interruptum Buckley, Proc. Acad. Phila. 100 (1862).
Plentiful but scattered, near the shores of Corpus Christi Bay.
March 21 (1464).

CHLORIS Swartz, Prod. Veg. Ind. Occ. 25 (1788).

Chloris cucullata Bisch. in Ann. Sc. Nat. (III.) 19 : 357 (1853).
Scattered in low, open land along Corpus Christi Bay; altitude, sea level to 35 feet.
March 14 (1449).
Chloris verticillata Nutt. Trans. Am. Phil. Soc. (II.) 5 : 150 (1837).
Plentiful in low, open pasture land near Gregory, San Patricio county, altitude 35 feet.
April 14 (1580).
On stony banks of the Guadalupe at Kerrville, Kerr county, altitude 1625 feet, but not plentiful.
May 19 (1767); type locality, Arkansas river.
Chloris verticillata intermedia Vasey, Cont. U. S. Nat. Herb. 2 : 528 (1894).
In a grassy meadow at the "Blind" Oso, nine miles southeast of Corpus Christi. Growing in rather thick clumps.
March 21 (1471).

BOUTELOUA Lag. in Varied. Cienc. (II.) 4 : 134 (1805).

Bouteloua hirsuta Lag. Varied. Cienc. 2 : Part 4, 141 (1805).

Chondrosium hirta H.B.K. Nov. Gen. 1 : 176 (1815).
On stony limestone hillsides in pasture land at Kerrville, Kerr county, altitude 1800 feet. Scattered, but rather plentiful.
June 18 (1878).

Bouteloua ramosa Scribn. Ill. N. A. Gr. 1 : Part 1, 44 (1890).
On the rocky left bank of the Guadalupe, at Kerrville, Kerr county, growing in scattered clumps, altitude 1630 feet.
May 19 (1762).

Bouteloua Texana S. Watson, Proc. Am. Acad. 18 : 196 (1883).
In a grassy meadow at the "Blind" Oso. Plentiful; seen also at Corpus Christi and at Kerrville.
March 21 (1485).

DACTYLOCTENIUM Willd. Enum. Hort. Berol. 1029 (1809).

Dactyloctenium Ægyptiacum (L.) Willd. Enum. Hort. Berol. 1029 (1809).
Cynosurus Ægyptiaca L. Sp. Pl. 72 (1753).
In cultivated ground at Corpus Christi, altitude 40 feet.
May 30 (1797); type locality, Africa.

LEPTOCHLOA Beauv. Agrost. 71, *t. 15, f. 1* (1812).

Leptochloa mucronata (Michx.) Kunth, Rev. Gram. 1 : 91 (1835).
Elusine mucronata Michx. Fl. Bor. Am. 1 : 65 (1803).
In loose rich ground on the banks of the Guadalupe, at Kerrville, altitude 1600 feet.
June 18 (1884); type locality, Illinois.

BULBILIS Raf. Am. Month. Mag. 190 (1819).

Bulbilis dactyloides (Nutt.) Raf. Am. Month. Mag. 190 (1819).
Sesleria dactyloides Nutt. Gen. 1 : 65 (1818).
Buchloë dactyloides Engelm. Trans. St. Louis Acad. Sci. 1 : 432 *t. 12* (1859).
Abundant about Corpus Christi in open pasture land, sea level to 40 feet.
March 21 (1447); type locality, plains of the Missouri.

SIEGLINGIA Bernh. Syst. Verz. Pfl. Erf 40 (1800).

[TRIODIA R. Br. Prodr. Fl. Nov. Holl. 1 : 182 (1810).]
[TRICUSPIS Beauv. Agrost. 77, *t. 15. f. 12.* (1812).]

Sieglingia acuminata (Munro) Kuntze, Rev. Gen. Pl. 789 (1891).

Tricuspis acuminata Munro; Vasey, Ill. N. A. Gr. 1 : Part 2, 32 (1891).
In dry, stony limestone ground at Kerrville, altitude 1650-1700 feet. Usually growing in bunches.
April 23 (1637).
Sieglingia congesta Dewey, Cont. U. S. Nat. Herb. 2 : 538 (1894).
Growing on the edge of depressions in the stiff, black "waxy land" between Corpus Christi and the Oso, altitude 35 feet. Prostrate, not plentiful.
March 21 (1486). From type locality. Type in U. S. Nat. Herb.

ERAGROSTIS Beauv. Agrost. 70, *t. 14, f. 11* (1812).

Eragrostis lugens Nees, Agrost. Bras. 507 (1829).
On the stony edge of the bluff overlooking the Guadalupe, at Kerrville. Only a few clumps seen, altitude 1650 feet.
May 14 (1745); type locality, South America.
Eragrostis capillaris (L.) Nees, Agrost. Bras. 505 (1829).
Poa capillaris L. Sp. Pl. 68 (1753).
On the edge of a water hole at Corpus Christi, altitude 40 feet. Scarce.
May 30 (1803); type locality, Virginia.
Eragrostis hypnoides (Lam.) B. S. P. Prel. Cat. N. Y. 69 (1888).
Poa hypnoides Lam. Tabl. Encycl. 1 : 185 (1791).
Poa reptans Michx. Fl. Bor. Am. 1 : 69 (1803).
Eragostis reptans Nees, Agrost. Bras. 514 (1829).
Creeping and forming matted masses in and on the edges of dried up water holes at Corpus Christi, altitude 40 feet.
March 17 (1455).
Eragrostis interrupta (Nutt.) Trelease.
Poa interrupta Nutt. Trans. Am. Phil. Soc. (II.) 5 : 146 (1837).
Eragrostis oxylepis Torr. Pac. R. R. Rep. (Whipple Exped.) 4 : 156 (1857).
A few scattered plants in open grassy pasture lands, near Gregory, San Patricio county, April 14 (1581), 35 feet, and on the edge of a water hole at Corpus Christi in black waxy land, altitude 40 feet.
May 30 (1802); type locality, Arkansas.
Eragrostis major Host. Gram. Austr. 4 : 14, *t. 24* (1809).
Briza Eragrostis L. Sp. Pl. 70 (1853).
Eragrostis poaeoides var. *megastachya* A. Gray, Man. Ed. 5, 631 (1867).
Eragrostis Eragrostis MacM. Met. Minn. 75 (1892), not Karst.

Prostrate, growing in yards in the shell deposit on the beach at Corpus Christi. May 30 (1783).

At Kerrville, on the banks of the Guadalupe; growing in the gravel near the water's edge, altitude 1600 feet, was a more robust form.

June 18 (1882); type locality, Europe.

Eragrostis Eragrostis (L.) Karst. Deutsche Fl. 389 (1880-83).
Poa Eragrostis L. Sp. Pl. 68 (1753).
Eragrostis poaeoides Beauv. Agrost. 162 (1812).
Eragrostis minor Host. Fl. Austr. 1: 135 (1827).

Growing among gravel along the Guadalupe at Kerrville; altitude 1600 feet. Scattered, but rather plentiful; only near the water's edge.

June 18 (1879); type locality, Italy.

MELICA L. Sp. Pl. 66 (1753).

Melica diffusa Pursh, Fl. Am. Sept. 77 (1814).
Melica altissima Walt. Fl. Car. 78 (1788), not L.
Melica mutica var. *diffusa* A. Gray, Man. Ed. 5, 626 (1867).

Growing in clumps on moist limestone ledges along the Guadalupe at Kerrville, altitude 1625 feet.

April 27 (1662); type locality, Virginia and Carolina.

UNIOLA, L. Sp. Pl. 71 (1753).

Uniola paniculata L. Sp. Pl. 71 (1753).
Uniola maritima Michx. Fl. Bor. Am. 1: 71 (1803).

On the Gulf coast of Mustang Island at Rope's Pass, growing in loose sand at sea level. Culms tall and stout.

May 28 (1783a); type locality, Carolina.

FESTUCA L. Sp. Pl. 73 (1753).

Festuca octoflora Walt. Fl. Car. 81 (1788).
Festuca bromoides Michx Fl. Bor Am. 1: 66 (1803).
Festuca tenella Willd. Enum. 1: 113 (1809).

A common grass in the region of Corpus Christi Bay, often growing in the sand along the Beach.

April 14 (1576); type locality, Carolina.

BROMUS L. Sp. Pl. 76 (1753).

[CERATACHLOA Beauv. Agrost. 75 (1812).]

Bromus unioloides (Willd.) H. B. K. Nov. Gen. 1: 151 (1815).
Festuca unioloides Willd. Hort. Berol. 1: 3, *t. 3* (1806).
Bromus Willdenowii Kunth, Rev. Gram 1: 134 (1829-35).
Bromus Schraderi Kunth, Enum. 1: 416 (1833).

Prostrate and spreading, in sand along and near the beach of Corpus Christi Bay. Seen at two stations, both at a distance from dwellings.
March 24 (1497).

ELYMUS L. Sp. Pl. 83 (1753).

Elymus Canadensis L. Sp. Pl. 83 (1753).
Elymus Philadelphicus L. Sp. Pl. Ed. 2, 122 (1762).
In rich, moist, shady ground on the banks of the San Antonio, at San Antonio, May 5 (1712). At Kerrville, on the stony limestone banks of the Guadalupe, altitude 1625 feet, occurs a stouter form.
May 19 (1763).

HORDEUM L. Sp. Pl. 84 (1753).

Hordeum pusillum Nutt. Gen. 1: 87 (1818).
Common in sand along the beach of Corpus Christi Bay, especially southeast of the town.
March 27 (1504).

CYPERACEAE.

CYPERUS L. Sp. Pl. 44 (1753).

Cyperus acuminatus Torr. & Hook. Ann. Lyc. N. Y. 3: 435 (1836).
In moist ground on the edge of a water hole at Corpus Christi, altitude 40 feet. Plants few and scattered.
May 30 (1807).

Cyperus aristatus Rottb. Descr. et Icon. 23, *t. 6, f. 1* (1773).
Cyperus inflexus Muhl. Gram. 16 (1817).
Cyperus uncinatus Pursh, Fl. Am. Sept. 50 (1814), not Poir.
In low, wet ground along the Guadalupe at Kerrville, altitude 1600 feet. Only a few plants seen.
June 28 (1924).

Cyperus ferax Vahl. Enum. 2: 357 (1806).
In moist gravel along the left bank of the Guadalupe at Kerrville, altitude 1600 feet. Usually freely branching from the base, forming tufts.
July 2 (1934).

ELEOCHARIS R. Br. Prodr. Fl. Nov. Holl. 1: 224 (1810).

Eleocharis capitata (L.) R. Br. Prodr. Fl. Nov. Holl. 1: 225 (1810).
Scirpus capitatus L. Sp. Pl. 48 (1753).
Eleocharis dispar E. J. Hill, Coult. Bot. Gaz. 7: 3 (1882).

On moist muddy rocks along the Guadalupe at Kerrville, altitude 1600 feet. Growing in round tufts, usually prostrate, but some of the larger plants ascending. Quite variable in size, the stems ranging from two to six or seven inches in length.

June 13 (1851).

DICHROMENA Michx. Fl. Bor. Am. 1: 37 (1803).

Dichromena colorata (L.) A. S. Hitchcock, 4th Rept. Mo. Bot. Gard. 141 (1893).
Schoenus coloratus L. Sp. Pl. 43 (1753).
Scirpus cephalotes Walt. Fl. Car. 71 (1788).
Dichromena leucocephala Michx. Fl. Bor. Am. 1 : 37 (1803).
Dichromena cephalotes Britton, Bull. Torr. Club. 15 : 100 (1888).

In stiff moist ground on the banks of the Guadalupe at Kerrville, growing in patches, altitude 1600 feet.

June 19 (1886).

Dichromena nivea Boeckl.
Dichromena Reverchoni S. H. Wright.

Growing in thick tufts in wet places on the left bank of the Gaudalupe, at Kerrville, altitude 1600 feet. Abundant at this one station, but not observed elsewhere. Lack of time has prevented the obtaining of the proper citations for this species.

May 23 (1778).

FUIRENA Rottb. Descr. et Ic. 70, *t. 19, f. 3* (1773).

Fuirena simplex Vahl. Enum. 2: 384 (1806).
Fuirena squarrosa Torr. Ann. Lyc. N. Y. 2: 252 (1828), not Michx.
Fuirena squarrosa var. *aristatula* Torr. Ann. Lyc. N. Y. 3: 291 (1836).

In wet gravel on the left bank of the Guadalupe, at Kerrville. Plants tall and stout, growing in clumps; altitude 1600 feet.

July 2 (1937).

CLADIUM P. Br. Civ. and Nat. Hist. Jam. 114 (1756).

Cladium effusum Torr. Ann. Lyc. N. Y. 3: 374 (1836).

In wet ground on both banks of the Guadalupe at Kerrville, but not abundant; altitude 1600 feet.

June 20 (1892).

BROMELIACEAE.

TILLANDSIA L. Sp. Pl. 286 (1753).*

Tillandsia recurvata L. Sp. Pl. Ed. 2; 410 (1762).

Collected on chapparral at Corpus Christi, where it is found occasionally. Near the mouth of the Nueces it is quite plentiful on trees, *Quercus*, *Celtis*, etc. Also observed on live oaks between Corpus Christi and Kenedy, and on trees at Kerrville, on the upper Guadalupe. Usually grows in ball-like masses.

March 6 (1400); type locality, Jamaica.

COMMELINACEAE.

COMMELINA L. Sp. Pl. 40 (1753).

Commelina Virginiana L. Sp. Pl. Ed. 2, 61 (1762).
Commelina angustifolia Michx. Fl. Bor. Am. 1: 24 (1803).

In sandy, cultivated fields at the Oso; flowers greenish blue. Seen also at Corpus Christi and Kerrville.

April 12 (1555).

TRADESCANTIA L. Sp. Pl. 288 (1753).

Tradescantia micrantha Torr. Mex. Bound. Surv. 2: 224 (1859).

Near the Oso, Nueces county, growing among clumps of low tangled bushes, altitude 25 feet. Owing to the difficulty in collecting it, only a few specimens were obtained, although it is rather plentiful. The flowers are small, rose color, apparently open only in the forenoon. The specimens were collected between 9 and 10 A. M., but when I returned in the afternoon not a single open flower could be seen.

April 12 (1564); type locality, near the mouth of the Rio Grande.

Tradescantia

On the plateau above Corpus Christi Bay, a short distance southeast of the town, altitude 40 feet, growing among chapparral, in rich black land. Flowers comparatively small, pale pink. This seems to be *Tradescantia leiandra* var. *brevifolia* Torr., and the var. (?) *ovata* of Coulter. It is certainly specifically distinct from *leiandra*, but the "caule

---

* *T. usneoides* L. reported in the Manual of Western Texas as occurring in Southern Texas, but not north of the mouth of the Pecos, was seen in abundance along the San Antonio and Aransas Pass R. R. quite a distance north of the Colorado. At the crossing of the Colorado it is found in great profusion.

prostrato" of Torrey's description does not fit, as this plant is ascending. It is the *Commelina speciosa* Buckley, Proc. Acad. Phila. 4 (1862), collected by Buckley at Corpus Christi.

March 10 (1447).

TINANTIA Scheidw Allgem. Gartenzeit. 7: 365 (1839).
Tinantia anomala (Torr.) Clarke.

Tradescantia anomala Torr. Méx. Bound. Surv. 2: 225 (1859).

In rich shady soil on Town Creek and along the Guadalupe at Kerrville, altitude 1625 feet. Stems erect, but weak, one to two feet high; flowers deep blue.

May 3 (1693); type locality, Austin and San Antonia.

JUNCACEAE.

JUNCUS L. Sp. Pl. 325 (1753).

Juncus filipendulus Buckley, Proc. Acad. Phila. 1862, 8 (1859).

Juncus leptocaulis T. & G. in Engelm. Trans. St. Louis Acad. 2: 454 (1866).

On moist rocks covered with a thin deposit of mud on the right bank of the Guadalupe, altitude 1600 feet. Plants scattered and not plentiful. Quite an extension of the range of this rather rare *Juncus*. "Apparently confined in Texas to the northern portion of the State."—Man. of W. Texas, 451.

June 13 (1852).

Juncus marginatus setosus Coville, Proc. Biol. Soc. Wash. 8: 124 (1893).

Very plentiful in dry soil at Corpus Christi, near the "Arroyo," altitude 40 feet. Not previously known to occur in the coast region.

May 30 (1796).

Juncus nodosus Texanus Engelm. Trans. St. Louis Acad. 2: 471 (1868).

Growing in dense matted clumps in mud on the left bank of the Guadalupe, at Kerrville, altitude 1600 feet. The roots were so interlaced and tangled that upon taking hold of a bunch a slight pull brought away a large section of the thin layer of mud, laying bare the rocks.

July 2 (1936).

Juncus tenuis Willd. Sp. Pl. 2: 214 (1799).

Two or three plants were found mixed with *J. filipendulus*. Apparently rare in the region of Kerrville, as no others were seen.

LILIACEAE.

ALLIUM L. Sp. Pl. 294 (1753).

Allium mutabile Michx. Fl Bor. Am. 1: 195 (1803).
In rich stony limestone soil along Bear Creek, Kerr county, altitude 1900 feet. Scape rather stout, flowers white.
April 30 (1684).

Allium Nuttallii S. Wats. Proc. Am. Acad. 14: 227 (1879).
Abundant in stony limestone ground about Kerrville, usually near bushes, altitude 1600–1800 feet. Found at the lowest elevations along the Guadalupe and on hillsides. Flowers rose color.
April 26 (1659).

NOTHOSCORDUM Kunth, Enum. 4:' 457 (1843).

Nothoscordum ornithogaloides (Walt.) Kunth, 4: 460 (1843).
Allium ornithogaloides Walt. Fl. Car. 121 (1788).
Allium striatum Jacq. Coll. Suppl. 51 (1796).
Nothoscordum striatum Kunth, Enum. 4: 459 (1843).
Common, but scattered, in low dry ground at Corpus Christi and surrounding country. Seen also at Kerrville.
March 10 (1397); type locality, Carolina.

YUCCA L. Sp. Pl. 319 (1753).

Yucca glauca stricta (Sims) Trelease, Fourth Ann. Rep. Mo. Bot. Gard. 206 (1893).
Yucca stricta Sims, Bot. Mag. *t. 2222* (1821).
Yucca angustifolia var. *mollis* Engelm. Trans. St. Louis Acad. 3: 50 (1873).
At two stations near Kerrville, on Town Creek, altitude 1650 feet. A low plant 15 inches to 3 feet high; leaves short, sparingly filamentous; scape rather densely flowered; flowers greenish white, tinged slightly with dull purple, especially on the outside. It had all the appearance of an introduced plant, growing only in cultivated fields.
May 3 (1689); type locality, Carolina.

Yucca rupicola Scheele, Linnæa, 23: 143 (1850).
On stony limestone hilltops near Kerrville, not common, altitude 1900 feet. The flowers on all the plants noticed were pure white, not greenish-white. Not seen in fruit.
May 21 (1775); type locality, New Braunfels.

Yucca Treculeana Carr. Rev. Hort. 7: 280 (1858).

Individual plants 6–10 feet high; occasional in the chapparral about Corpus Christi. Leaves dark green, sparingly filamentous near the base. A species of light brown wasp, flies, and bugs were observed on and in the flowers. Through carelessness, the leaves became spoiled and unfit for specimens.

DASYLIRION Zucc. in Otto & Dietr. Allg. Gartenz. 6: 258 (1838).

Dasylirion Texanum Scheele, Linnæa, 23: 140 (1850).

A single plant found in bloom near the summit of a conical terraced hill at Kerrville, altitude 1900 feet.

June 30 (1929); type locality, New Braunfels.

SCHOENOCAULON A. Gray, Ann. Lyc. Nat. Hist. N. Y. 4: 127 (1837).

Schoenocaulon Texanum Scheele, 25: 262 (1852).

Schoenocaulon Drummondii A. Gray, Bot. Beechy, 388 (1841), name only.

Very common on the dry limestone hills about Kerrville, ranging from the lowest elevations along the Guadalupe and Town Creek to near the summits of the hills, 1600–1900 feet.

April 25 (1626); type locality, New Braunfels.

SMILACEAE.

SMILAX L. Sp. Pl. 1028 (1753).

Smilax Bona-Nox L. Sp. Pl. 1030 (1753).

Smilax tamnoides A. Gray, Man. 485 (1848), not L.

Climbing over bushes along Town Creek, altitude 1650 feet.

April 28 (1674).

Smilax rotundifolia L. Sp. Pl. 1030 (1753).

Smilax caduca L. Sp. Pl. 1030 (1753).

Climbing high over bushes and trees along Bear Creek, Kerr county, altitude 1800 feet.

April 30 (1679).

AMARYLLIDACEAE.

COOPERIA Herb. Bot. Reg. *t. 1835* (1836).

Cooperia Drummondii Herb. Bot. Reg. *t. 1835* (1836).

In dry open ground at Corpus Christi, altitude 10 feet. Flowers

white, tinged with dull purple on the outside. Observed in dry ground along the Guadalupe near Kerrville.
June 6 (1826).

Cooperia pedunculata Herb. Amaryll. 179, *t. 42, f. 3-5* (1837).
On stony slopes along the Guadalupe, usually in rich shaded ground, altitude 1625 feet. Outside of perianth more purple marked than *C. Drummondii*, the tube much longer.
April 19 (1611).

AGAVE L. Sp. Pl. 323 (1753).

Agave maculata Regel, Ind. Sem. Hort. Petrop. 16: (1856).
Common about Corpus Christi. Flowers purplish green on all the specimens noticed. One of the "rattlesnake plants." The root or rather crown is said to be an antidote for snake bites.
June 2 (1815).

IRIDACEAE.

CALYDOREA Herb. in Lindl. Bot. Reg. Misc. 85 (1843).

Calydorea Texana (Herb.) Baker, Journ. Bot. 14: 189 (1876).
Gelasine? Texana Herb. Bot. Mag. *t. 3779* (1840).
Collected by Drummond, Galveston Bay, and apparently not since found. Scattered, but not uncommon in hard dry ground about Corpus Christi and along Nueces Bay. The showy blue flowers are very delicate, and although a number of plants were collected, only four or five were fit for specimens by the time they were brought in.
March 8 (1403); type locality, Texas.

SISYRINCHIUM L. Sp. Pl. 954 (1753).

Sisyrinchium Bermudianum L. Sp. Pl. 954 (1753).
Sisyrinchium angustifolium Mill. Dict. Ed. 8 (1768).
Sisyrinchium anceps Cav. Diss. 6: 345, *t. 190, f. 2* (1788).
Sisyrinchium gramineum Curtis, Bot. Mag *t. 464* (1799).
Sisyrinchium mucronatum Michx. Fl. Bor. Am. 2: 33 (1803).
In sand near the Oso, growing in tufts, flowers large. Sea level.
April 12 (1552).

CANNACEAE.

CANNA L. Sp. Pl. 1 (1753).

Canna Indica L. Sp. Pl. 1 (1753).
In low ground along the San Antonio at San Antonio. Stems 5-6 feet high; altitude, 600 feet.
June 9 (1839); type locality, tropics of both continents.

JUGLANDACEAE.

JUGLANS L. Sp. Pl. 997 (1753).

Juglans nigra L. Sp. Pl. 997 (1753).
A few trees were noticed at San Antonio near the S. P. bridge, altitude 600 feet.

Juglans rupestris Engelm. in Torr. Sitgr. Rep. 171, *t. 15* (1853).
A bush or small spreading tree, very common in low ground along the Guadalupe, altitude 1600 feet.
April 19 flower, July 2 fruit (1615); type locality, New Mexico.

HICORIA Raf. Med. Rep. (II.) 5: 352 (1808).
[HICORIUS Raf. Fl. Lud. 109 (1817).]
[CARYA Nutt. Gen. 2: 221 (1818).]

Hicoria Pecan (Marsh.) Britton, Bull. Torr. Club, **15**: 282 (1888).
Juglans Pecan Marsh. Arb. Am. 69 (1785).
Juglans Illinoiensis Wang. Beitr. Holz. Am. 54, *t. 18, f. 43* (1787).
Carya olivaeformis Nutt. Gen. 2: 221 (1818).
In a wooded tract along the S. P. R. R. at San Antonio, several large trees were seen in flower, altitude 600 feet.
April 17 (1589).

SALICACEAE.

SALIX L. Sp. Pl. 1015 (1753).

Salix nigra Marsh. Arb. Am. 139 (1785).
Occurring as a bush, or slender tree 25 feet high, along the Guadalupe, altitude 1600 feet. Typical.
April 19 (1621).

Salix nigra longipes forma **venulosa** And. Monog. Sal. 22 (1867).*

---

*Cont. U. S. Nat. Herb. 4: 199 (1893), as *S. nigra venulosa*.

Two forms of this willow were collected, one in flower (1643), April 24, about a mile and a half below Kerrville, near the mouth of a tributary of the Guadalupe, which empties from the right bank. The specimens were from a slender bush or clump of bushes about 10 feet high. The leaves are comparatively short and broad in proportion, very white underneath. The other, collected on the right bank of the Guadalupe, about a mile below Kerrville, was a branching bush 10–15 feet high, in fruit (1902), June 22. The leaves are longer and narrower, more like *S. nigra*, light green, shining, and less whitened beneath; altitude 1600 feet.

Type locality, "in Nova Mexico."

FAGACEAE.

QUERCUS L. Sp. Pl. 994 (1753).

Quercus cinerea Michx. Fl. Bor. Am. 197 (1803).

A bush about 7 feet high on the left bank of the Guadalupe, at Kerrville, on a moist limestone ledge. The leaves seem to be deciduous, as there were no signs of old ones. Apparently not recorded from so far south and west, its range given as "sandy barrens, extending from the Gulf States to the valley of the Brazos."

April 19 (1616); type locality, Carolina and Georgia.

Quercus coccinea Wang. Am. 44, *t. 4, f. 9* (1789).

Quercus rubra var. β .L. Sp. Pl. Ed. 2, 1414 (1763).

Occurring as a small spreading tree along the Guadalupe and its tributaries about Kerrville, altitude 1650 feet.

April 19 (1639).

Quercus Virginiana Mill. Gard. Dict. Ed. 8, No. 16 (1768).

Quercus virens Ait. Hort. Kew. 3: 356 (1789).

Occurring as straggling bushes 6–8 feet high at Flower Bluff, near the Gulf coast, altitude 15 feet.

April 9 (1542).

ULMACEAE.

CELTIS L. Sp. Pl. 1043 (1753).

Celtis occidentalis L. Sp. Pl. 1044 (1753).

A small tree in the region of Corpus Christi, along Nueces Bay, at sea level. At San Antonio it often occurs as a large spreading tree with rough, corky bark and thick coriaceous leaves, whitened beneath.

April 17 (1587); type locality, Virginia.

Celtis Mississippiensis Bosc. Encycl. Agric. 7: 577 (1822).

Growing in company with *C. occidentalis* at San Antonio. Usually a smaller tree, with smoother bark, and thin, light green leaves. The fruit is bright light brown, smooth and clear-looking; altitude 600 feet. April 17 (1586).

Celtis Tala Gill.; Planch. Ann. Sc. Nat. (III.) 10: 310 (1848).

Celtis pallida Torr. Mex. Bound. Surv. 2: 203 (1859).

Very common in the coast region about Corpus Christi, altitude 10-40 feet. Usually a thick, spreading flexuous bush, but occasionally slender and tree-like.

Collected near Gregory, San Patricio county, April 14 in flower, and at Corpus Christi June 8 in fruit.

(1570); type locality, South America.

MORACEAE.

MORUS L. Sp. Pl. 986 (1753).

Morus rubra L. Sp. Pl. 986 (1753).

A small tree in rich shaded ground on the left bank of the Guadalupe at Kerrville, altitude 1610 feet. A southern and western extension of the range, which is given as "extending to the valley of the Colorado, in Texas."

April 19 (1605); type locality, Virginia.

Morus nigra L. Sp. Pl. 986 (1753).

Rather common in cultivation at Corpus Christi. A single tree found growing wild near the beach, along the upper part of the Bay. Fruit black, sour.

March 14 (1448); type locality, Italy.

URTICACEAE.

URTICA L. Sp. Pl. 984 (1753).

Urtica urens L. Sp. Pl. 984 (1753).

A common weed in yards, waste places, and along the streets at Corpus Christi from sea level to 40 feet. Usually low and branching from the decumbent base. The effects of contact with the stinging hairs of this plant last about 18 or 20 hours.

March 10 (1393); type locality, Europe.

BOEHMERIA Jacq. Stirp. Am. 246, *t. 157* (1763).
Boehmeria cylindrica (L.) Willd. Sp. Pl. 4: 340 (1809).
Urtica cylindrica L. Sp. Pl. Ed. 2, 1396 (1763).
In rich, damp, shady ground, along the left bank of the Guadalupe, above Kerrville, altitude 1600 feet.
June 21 (1900); type locality, Jamaica.

PARIETARIA L. Sp. Pl. 1052 (1753).
Parietaria debilis Forst. f. Prodr. 73 (1786).
Abundant about Corpus Christi, from sea level to 35 feet. In open and exposed places flat and prostrate; becoming ascending with long weak stems in rich shaded ground.
March 24 (1499); type locality, tropics.

LORANTHACEAE.

PHORADENDRON Nutt. Journ. Phila. Acad. (II.) 1: 185 (1847-50).
Phoradendron flavescens (Pursh) Nutt.; A. Gray. Man. Ed. 2, 383 (1856).
Viscum flavescens Pursh, Fl. Am. Sept. 114 (1814).
Very common about Waco, McLennan county, on *Prosopis juliflora* and other trees. Seen at various places between Waco and Corpus Christi on *Quercus Virginiana*, and along Nueces Bay on *Celtis occidentalis*. Apparently not reported before from extreme Southern Texas, its southernmost range being given as "from Eagle Pass to Central Texas."
March 2 (1376); type locality not given.

POLYGONACEAE.

ERIOGONUM Michx. Fl. Bor. Am. 1: 246 (1803).
Eriogonum longifolium Nutt. Trans. Am. Phil. Soc. (II.) 5: 164 (1833-37).
Eriogonum Texanum Scheele, Linnæa, 22: 150 (1849).
Occurring in limestone soil about Kerrville from the lowest elevations along the Guadalupe to the summit of the highest hills, where it was most plentiful; altitude 1625 to 2000 feet.
June 18 (1877); type locality, Arkansas.

RUMEX L. Sp. Pl. 333 (1753).

Rumex spiralis Small, Bull. Torr. Club, **22**: 44 (1895). *

[PLATE 1.]

Perennial, slender, glabrous, light green, somewhat glaucescent. Rootstock woody, creeping, 1-2 dm. long; roots fibrous; stem erect, 8-9 dm. long, simple or sparingly branched above, leafy throughout, slightly flexuous, strongly channeled, woody below; leaves lanceolate or oblong-lanceolate, 6-13 cm. long, 1.5-4.5 cm. broad, acute or sometimes attenuate at the apex, the lower ones obtuse or truncate at the base, the upper acute or acuminate at the base, all rather long petioled, coriaceous, light-green, undulate and crisped, neither prominently nor conspicuously nerved; petioles strict, 2-5 cm. long; ocreae cylindric, nearly one-half as long as the internodes; inflorescence terminal, simply paniculate, naked; racemes (fruiting) 5-12 cm. long, dense, rather erect, the terminal one usually about twice as long as the lateral ones; calyx 2 mm. broad; pedicels varying from 2-4 mm. in length, jointed below the middle; wings broadly ovate cordate, broader than high, 1 cm. long, 1-1.2 cm. broad, straw-colored, sometimes slightly constricted below the apex, conspicuously and prominently nerved, crenulate and undulate, each one bearing an oblong-ovoid callosity, the three wings strongly spirally twisted; achene broadly oblong-ovoid, 3 mm. long, short-pointed, chestnut-colored, its faces nearly flat, its angles conspicuously margined.

Found growing in the mud on the margins of ponds near Kenedy, Carnes county, Texas, by Mr. A. A. Heller, collected in flower and fruit on May 26, 1894 (1781). The altitude of the station is about 400 feet.

Its nearest relative is *Rumex altissimus*, from which, however, it differs in having more characteristically lanceolate leaves, which are longer-petioled, crisped and the larger ones more or less truncate at the base instead of acuminate. The panicle of *R. spiralis* is more open, not leafy, and its racemes are denser and thicker. Wings twice to thrice as large as in *R. altissimus* invest the broadly oblong-ovoid achene. The former are broader than high and strikingly cordate, whereas those of *R. altissimus* are higher than broad, not strongly cordate and less prominently nerved. So far as observed three callosities are developed throughout.

POLYGONUM L. Sp. Pl. 359 (1753).

Polygonum densiflorum Meisn. in Mart. Fl. Bras. **5**: Part 1, 13 (1855).

---

* Description and plate reproduced by Mr. Small's permission.

Collected in mud and water in the San Antonio river, at the S. P. bridge, in mature fruit, May 3 (1836), 600 feet.

On July 2 it was collected in a similar situation on the Guadalupe, at Kerrville, in flower only (1942), 1600 feet.

Type locality, "Banda orientale Brasiliae." Type in Columbia College Herbarium.

Polygonum lapathifolium L. Sp. Pl. 360 (1753).

Polygonum Pennsylvanicum Curt. Fl. Lond. *t. 73* (1777), not L.

At Kerrville, on the edge of the bluff overlooking the Guadalupe, in wet ground, altitude 1650 feet, growing in a dense clump. Leaves very resinous beneath; flowers greenish white, dense.

June 12 (1844); type locality, "in Gallia."

CHENOPODIACEAE.

CHENOPODIUM L. Sp. Pl. 218 (1753).

Chenopodium album L. Sp. Pl. 219 (1753).

About the streets of Kerrville, apparently not very abundant. Altitude 1650 feet.

June 30 (1928); type locality, "in agris Europae."

ATRIPLEX L. Sp. Pl. 1052 (1753).

Atriplex

In sandy soil along the beach of Corpus Christi Bay. Seen only in flower.

June 2 (1819).

SUAEDA Forsk. Fl. Aeg. Arab. 69, *t. 186* (1775).

Suaeda suffrutescens S. Wats. Proc. Am. Acad. 9: 88 (1874).

Suaeda fruticosa var.? *multiflora* Torr. Pac. R. R. Rep. 4: 130 (1857), in part.

Abundant on the "Flats" at Corpus Christi, growing in dense bunches. Sea level.

June 6 (1827); type locality not given; range, "from western Texas to southern California and northern Mexico, in saline plains."

AMARANTACEAE.

Amaranthus Berlandieri (Moq.) Uline & Bray, Bot. Gaz. 19: 268 (1094).

Sarratia Berlandieri Moq.; DC. Prodr. 132: 268 (1849).

About Corpus Christi, in both sandy soil and black waxy land, from sea level to 40 feet, and at Kerrville on the left bank of the Guadalupe, on rocks thinly covered with soil, 1600 feet.

April–June (1487); type locality, Mexico.

Amaranthus blitoides S. Wats. Proc. Am. Acad. 12 : 273 (1876).

Very abundant about Kerrville along gutters and in open lots and yards. Altitude, 1650 feet.

June 15 (1867); type locality not given, but "frequent in the valleys and plains of the interior, from Mexico to N. Nevada and Iowa."

Amaranthus Palmeri S. Wats. Proc. Am. Acad. 12 : 274 (1876).

On the left bank of the Guadalupe at Kerrville in both shaded and open places but not plentiful. Altitude 1625 feet.

June 19 (1890); type locality, "at Larkin's Station, San Diego county, California."

Amaranthus retroflexus L. Sp. Pl. 991 (1753).

Common in rich ground about the streets of Kerrville, altitude 1650 feet.

May 19 (1765); type locality, "in Pennsylvania."

CLADOTHRIX Nutt.; Moq. in DC. Prodr. 13: Part 2, 359 (1849).

Cladothrix lanuginosa Nutt.; Moq. in DC. Prodr. 13: Part 2, 360 (1849).

Achryanthes lanuginosa Nutt. Trans. Am. Phil. Soc. (II.) 5 : 166 (1833–37).

Prostrate in the sand along the beach of Corpus Christi Bay. Much branched and spreading.

June 2 (1813); type locality, "secus Salt River et Red River."

ALTERNANTHERA Forsk. Fl. Aegypt. Arab. 28 (1775).

Alternanthera repens (L.) Kuntze, Rev. Gen. Pl. 540 (1891).

Achyranthes repens L. Sp. Pl. 205 (1753).

Illecebrum Achyrantha L. Sp. Pl. Ed. 2, 299 (1763).

Alternanthera Achyrantha R. Br. Prodr. Fl. Nov. Holl. 1: 417 (1810).

A common weed in yards and open places in rich ground at Kerrville, altitude 1650 feet.

June 15 (1896); type locality, "in Turcomannia."

GOMPHRENA L. Sp. Pl. 224 (1753).

Gomphrena Neallyi Coult. & Fisher, Cont. U. S. Nat. Herb. 2: 363 (1894).

Gomphrena nitida Coulter, Cont. U. S. Nat. Herb. No. 2, 48 (1890), not Rothrock.

Rather plentiful about Corpus Christi, in dry ground, often freely branching from the large fusiform root, thus making a clump of a dozen or more stems. In March it was first found in flower when only two or three inches high, and in June plants from 1-2 feet high were collected; altitude, sea level to 40 feet.
March 8 (1408); from type locality.

IRESINE P. Br. Civ. & Nat. Hist. Jam. 358 (1755).

Iresine vermicularis (L.) Moq. in DC. Prod. 13: 2, 340 (1849).
Gomphrena vermicularis L. Sp. Pl. 224 (1753).
Illecebrium vermiculatum L. Sp. Pl. Ed. 2, 300 (1762).
Philoxerus vermiculatus Rees. Cycl. 5: 27 ( ).
Growing in dense tangled bunches on the flat, sandy shore of Mustang Island, on the west side, at Rope's Pass. Distributed as *Philoxerus vermicularis* Moq.
May 28 (1784); type locality, "in Brasilia, Curassao."

PHYTOLACCACEAE.

RIVINA L. Sp. Pl. 121 (1753).

Rivina humilis L. Sp. Pl. 121 (1753).
Growing by and under bushes at Corpus Christi; also at Kerrville.
March 9 (1422), sea level to 35 feet at Corpus Christi; 1625 feet at Kerrville.

PHYTOLACCA L. Sp. Pl. 441 (1753).

Phytolacca decandra L. Sp. Pl. Ed. 2, 631 (1762).
In rich, shaded ground, on the left banks of the Guadalupe, at Kerrville, altitude 1620 feet.
June 19 (1891); type locality, "in Virginia."

BATIDEAE.

BATIS P. Br. Civ. & Nat. Hist. Jam. 358 (1755).

Batis maritima L. Sp. Pl. Ed. 2, 1451 (1763).
Growing luxuriantly on the "Flats" at Corpus Christi, especially in the moister portions. Sea level.
June 6 (1825); type locality, "in Jamaicae maritimis salsis."

NYCTAGINACEAE.

ALLIONIA Loefl. Iter Hisp. 181 (1758).

[Oxybaphus L'Her; Willd. Sp. Pl. 1: 185 (1797).]

Allionia albida Walt. Fl. Car. 84 (1788).
Calymenia albida Nutt. Gen. 1: 26 (1818).
Oxybaphus albidus Choisy in DC. Prodr. 13: Part 2, 433 (1849).

Occurring on the grassy plateau along Corpus Christi Bay, in rich black land; altitude 35 feet.

April 9 (1545); type locality, Carolina.

Allionia nyctaginea Michx. Fl. Bor. Am. 1: 100 (1803).
Calymenia nyctaginea Nutt. Gen. 1: 26 (1818).
Oxybaphus nyctagineus Sweet, Hort. Brit. 429 (1830).

On the left bank of Town Creek, at Kerrville, in rich shaded ground, altitude 1625 feet.

May 17 (1757); type locality, "ad ripas fluminis Tennessee."

MIRABILIS L. Sp. Pl. 177 (1753).

Mirabilis Jalapa L. Sp. Pl. 177 (1753).

Two or three large spreading plants were growing on the bluff along the Guadalupe at Kerrville. Flowers rose color.

June 18 (1881); type locality, India.

BOERHAVIA L. Sp. Pl. 3 (1753).

Boerhavia linearifolia A. Gray, Am. Journ. Sci. (II.) 15: 322 (1853).

Found sparingly in gravel on the right bank of the Guadalupe at Kerrville, altitude 1600 feet.

June 13 (1849); type locality, W. Texas.

Boerhavia obtusifolia Lam. Ill. 1: 10 (1791).
Boerhavia viscosa Lag. & Rodr. Anal. Cienc. Nat. 4: 256, No. 12 (1801).
Boerhavia patula Domb.; Lag. Enum. 1: 287 (1805).

Open ground at Corpus Christi, where it is plentiful; altitude, sea level. Flat on the ground, except the ends of the branches, which are slightly ascending. The small flowers are dark rose purple in color. Pena, Duval county, seems to have been its previous eastern limit.

May 29 (1792); type locality, Central America.

AIZOACEAE.

MOLLUGO L. Sp. Pl. 89 (1753).

Mollugo verticillata L. Sp. Pl. 89 (1753).
In cultivated ground at Corpus Christi, altitude 40 feet.
May 30 (1798); type locality, "in Africa, Virginia."

SESUVIUM L. Syst. Ed. 10, 1058 (1759).

Sesuvium Portulacastrum L. Sp. Pl. Ed. 2, 684 (1762).
Portulaca Portulacastrum, L. Sp. Pl. 446 (1753).
? *Polecarpon uniflorum* Walt. Fl. Car. 83 (1788).
Sesuvium pedunculatum Pers. Syn. Pl. 2 : 39 (1807).
Aizoon Canariense Andr. Bot. Rep. *t. 201.*
In moist saline soil at the Oso, 9 miles southeast of Corpus Christi.
April 9 (1534); type locality, "in Indiae maritimus."

PORTULACACEAE.

TALINUM Adans. Fam. Pl. 2: 245 (1765).

Talinum lineare H.B.K. Nov. Gen. 6: 77 (1823).
Talinum aurantiacum Engelm. Bost. Jour. Nat. Hist. 6: 153 (1850).
In dry, open pasture land near Gregory, San Patricio county, altitude 35 feet. Less than a half dozen plants were seen.
April 14 (1568); type locality, "locis aridis, inter Mexico et Real de Pachuca, prope Gasave, in valli Tenochtitlanensi?"

CARYOPHYLLACEAE.

SILENE L. Sp. Pl. 416 (1753).

Silene antirrhina L. Sp. Pl. 419 (1753).
In gutters and waste places about the streets of Kerrville, growing in patches, but not yet abundant; altitude 1650 feet.
June 15 (1865); type locality, "in Virginia, Carolina."

ARENARIA L. Sp. Pl. 423 (1753).

Arenaria Benthami Fenzl.; T. & G. Fl. N. A. 1: 675 (1840).
Arenaria monticola Buckley, Proc. Acad. Phila. 449 (1861) fide
  Gray same 161 (1862).
On a rocky limestone ridge along Wolf Creek, northeastern part of

Kerr county, growing in situations similar to Buckley's plants; altitude 1800 feet.

April 30 (1676); type locality, "Texas," collected by Drummond.

TISSA Adans. Fam. Pl. 2: 507 (1763).

Tissa diandra (Guss.) Britton, Bull. Torr. Club, 16: 128 (1889).
Arenaria diandra Guss. Fl. Sic. Prodr. 1: 515 (1827).
Arenaria salsuginea Bunge in Ledeb. Fl. Alt. 2: 163 (1829).
Spergularia diandra Boiss. Fl. Orient. 1: 733 (1867).

In sand at Corpus Christi, covered by sea water during storms when the water is forced back over the beach. Also in a depression not affected by salt water, but where water collects during rains, and in dry ground within the enclosure of the "Bluff City Park."

March 9 (1413).

PARONYCHIA Adans. Fam. Pl. 2: 272 (1763).

Paronychia setacea T. & G. Fl. N. A. 1: 170 (1838).

Abundant on the summits of the limestone hills about Kerrville. Also seen in the low ground along Town Creek, altitude 1625–2000 feet.

May 14 (1729); type locality, "Texas," collected by Drummond.

RANUNCULACEAE.

DELPHINIUM L. Sp. Pl. 530 (1753).

Delphinium Carolinianum Walt. Fl. Car. 155 (1788).
Delphinium azureum Michx. Fl. Bor. Am. 1: 314 (1803).

At San Antonio, on the right bank of the San Antonio, near the S. P. bridge, occurred a white flowered form, or sometimes tinged with pink. April 17 (1583), 600 feet.

On the hillsides and summits about Kerrville it was never white, but bright blue or pinkish. May 8 (1723), 1700–2000 feet.

Type locality not given by Walter; "in Carolina et Georgia," by Michx.

CLEMATIS L. Sp. Pl. 543 (1753).

Clematis Simsii Sweet, Hort. Brit. 1: 1 (1827).
Clematis cordata Sims, Bot. Mag. *t. 1816* (1816), not Pursh.
Clematis Pitcheri T. & G. Fl. N. A. 1: 10 (1838).

Frequent along the Guadalupe and Town Creek, at Kerrville, climbing over bushes; altitude 1600 feet.

April–June (1607).

Clematis Texensis Buckley, Proc. Phila. Acad. 448 (1861).
Occurring sparingly along the Guadalupe and Town Creek, at Kerrville. This plant agrees very well with Buckley's description, the "caule scandente, foliis pedunculatis, integris, lato-ovatis, acuminatis mucronatis," being quite correct, but the thin leaves are coriaceous, which character is much more apparent in dried specimens than in fresh ones. They are light green, glaucous beneath. "Part of these stem leaflets are on long tendril-like petioles," corresponds also. The slender, conical calyx is slightly contracted near the middle, an inch or more long, bright scarlet in color, rarely reflexed at the tips and then only slightly.

The figure in Bot. Mag. *t. 6594*, of *C. coccinea* Engelm. does not agree with this plant, neither do specimens in the Columbia College herbarium, cultivated at Easton, Pa. *C. coccinea* has much shorter and broader flowers, thick, coriaceous, and rounder and more recticulated leaflets. Distributed as *C. "Texana"* Buckley. A form intermediate between this plant and *C. Simsii* was collected on the Guadalupe, in close proximity to both. The leaflets are more coriaceous than those of either of the others, glaucous, though less so than in *C. Texensis.* The flower is dull purple red, or rather brick red, and in shape more like that of *C. Simsii.*

April 19 to June 26 (1608); type locality, "on the Colorado river, above Austin."

RANUNCULUS L. Sp Pl. 548 (1753).

Ranunculus macranthus Scheele, Linnaea 21 : 585 (1848).
Ranunculus repens, var *macranthus* A. Gray, Bost. Jour. Nat. Hist. 6: 141 (1850).
In a wet place on the stony limestone plateau about six miles northeast of Kerrville, altitude 1900 feet.
April 30 (1688); type locality, " prope Neu Braunfels."
Ranunculus trachyspermus Engelm. Bost. Jour. Nat. Hist. 5 : 211 (1847).
Growing in damp rich "black waxy land" at Corpus Christi, at an altitude of about 40 feet. Very scarce.
March 17 (1457); type locality, " Margins of ponds near Houston."

BERBERIDACEAE.

BERBERIS L. Sp. Pl. 330 (1753).
Berberis trifoliolata Moric. Pl. Nouv. Am. 113, *t. 69* (1833-46).

Berberis trifoliata Hartw.; Lindl. Bot. Reg. **27**: Misc. 68–31, *t. 10 ( )*.

Abundant about Corpus Christi, ranging from sea level along Nueces Bay to 40 feet on the plateau. Distributed as *B. "trifoliata,"* Moric. March 6 (1384).

MENISPERMACEAE.

CEBATHA Forsk. Fl. Ægypt. 171 (1775).
[COCCULUS DC. Syst. Veg. **1**: 515 (1818).]

Cebath Carolina (L.) Britton, Mem. Torr. Club, **5**: 162 (1894).
Menispermum Carolinum L. Sp. Pl. 340 (1753).
Cocculus Carolinus DC. Syst. Veg. **1**: 524 (1818).
Cebatha Virginica Kuntze, Rev. Gen. Pl. 9 (1891).

In dry open ground at Kerrville, at an altitude of 1650 feet. Plants less than a foot long, growing in bunches.

June 26 (1915); type locality, "in Carolina."

PAPAVERACEAE.

ARGEMONE L. Sp. Pl. 508 (1753).

Argemone Mexicana L. Sp. Pl. 508 (1753).

At Corpus Christi from sea level to 40 feet, but most abundant along the S. A. & A. P. Railroad, the embankment being white with it. No yellow-flowered ones were seen, nor any of the *A. platyceras rosea* Coulter, the type of which was collected here.

March 5 (1378); type locality, "in Mexico, Jamaica, Carabaeis."

CAPNOIDES Adans. Fam. Pl. **2**: 431 (1763).
[NECKERIA Scop. Introd. 313 (1777).]
[CORYDALIS Vent. Choix. 19 (1803).]

Capnoides micranthum (Engelm.) Britton, Mem. Torr. Club, **5**: 166 (1894).
Corydalis aurea var. *micrantha* Engelm. in A. Gray, Man. Ed. 5, 62 (1867).
Corydalis micrantha A. Gray, Coult. Bot. Gaz. **11**: 189 (1886).

First collected along Nueces Bay at sea level, as a small upright plant, and later at Corpus Christi at 35 feet, occurring as a weak slender reclining plant two feet high. Both stations were in rich shaded ground. Scarce. Distributed as *C. aureum.*

March 12 (1433); type locality, "W. Illinois and St. Louis."

CRUCIFERAE.

LEPIDIUM L. Sp. Pl. 642 (1753).

Lepidium intermedium A. Gray, Pl. Wright. 2: 15 (1853).

At Corpus Christi within the enclosure of the "Bluff City Park," at sea level, where it is plentiful.

March 9 (1421); type locality, "ravines of the Organ Mountains, northeast of El Paso."

Lepidium lasiocarpum tenuipes S. Wats. Proc. Am. Acad. 17: 322 (1882).

Growing in thick patches, several of which occur at Kerrville, in rich open ground in the town, altitude 1650 feet. Plant clammy puberulent, pod with hispid scattered hairs.

April 25 (1651).

Lepidium Virginicum L. Sp. Pl. 645 (1753).

In sand at Corpus Christi at sea level. A low procumbent bushy form.

March 24 (1495); type locality, "in Virginia, Jamaicae glareosis."

SISYMBRIUM L. Sp. Pl. 657 (1753).

Sisymbrium pinnatum (Walt.) Greene, Bull. Cal. Acad. 2: 390 (1887).

Erysimum pinnatum Walt. Fl. Car. 174 (1788).

Cardamine (?) *multifida* Pursh, Fl. Am. Sept. 440 (1814).

Sisymbrium canescens Nutt. Gen. 2: 68 (1818).

In sandy soil at the Oso, under bushes, at about 15 feet altitude. Also plentiful in yards at Corpus Christi.

March 21 (1470); type locality not given.

RORIPA Scop. Fl. Carn. 520 (1760).

[NASTURTIUM R. Br. in Ait. Hort. Kew. Ed. 2, 4: 109 (1812).]

Roripa Nasturtium (L.) Rusby, Mem. Torr. Club, 3: No. 3, 5 (1893).

Sisymbrium Nasturtium L. Sp. Pl. 657 (1753).

Nasturtium officinale R. Br. in Ait. Hort. Kew. Ed. 2, 4: 110 (1812).

Common at Kerrville in wet places, in a little stream in the town, and on the Guadalupe, from 1600–1650 feet altitude.

May 16 (1753); type locality, "in Europa and America septentrionali ad fontes."

Roripa tanacetifolia (Walt.).
Sisymbrium tanacetifolium Walt. Fl. Car. 174 (1788).
Nasturtium tanacetifolium Hook. & Arn., in Hook. Journ. of Bot. 1: 190 (1834).
At Corpus Christi on the edge of dried-up water holes, altitude 35 feet. Very plentiful near the " Arroyo."
March 23 (1488); type locality, Carolina.

LESQUERELLA S. Wats. Proc. Am. Acad. 23: 249 (1888).
Lesquerella Gordoni (A. Gray) S. Wats. Proc. Am. Acad. 23: 253 (1888).
Vesicaria Gordoni A. Gray, Bost. Jour. Nat. Hist. 6: 148 (1850).
Vesicaria angustifolia A. Gray, Pl. Wright, 2: 13 (1853), not Nutt.
Plentiful on the grassy plateau southeast of Corpus Christi, in rich black land, at an altitude of 15-35 feet.
March 21 (1478); type locality, "on the Canadian, in the Raton Mountains."
Lesquerella recurvata (Engelm.) S. Wats. Proc. Am. Acad. 23: 253 (1888).
Vesicaria recurvata Engelm.; Gray, Bost. Jour. Nat. Hist. 6: 147 (1850).
Vesicaria angustifolia Scheele, Linnaea, 21: 584 (1848), not Nutt.
Abundant on the summits of the hills about Kerrville, altitude 2000 feet. Also observed along the Guadalupe, altitude 1600 feet.
April 26 (1657); type locality, "San Antonio and New Braunfels."

DRABA L. Sp. Pl. 642 (1753).
Draba brachycarpa Nutt.; T. & G. Fl. N. A. 1: 108 (1838).
In open ground about Waco, McLennan county, altitude 400 feet, where it is plentiful. At that time only an inch or two high. Not recorded in the Manual of Western Texas.
March 2 (1371); type localities, St. Louis, Milledgeville and Macon, Georgia.
Draba cuneifolia Nutt.; T. & G. Fl. N. A. 1: 108 (1838).
Collected first in fields at Waco, McLennan county, altitude 400 feet, later at Corpus Christi, altitude about 6 feet, on the shell deposit in the yard of Ritter's Hotel. This latter form (1379) is very low and stunted, but with large pods. Later it was collected on limestone hillsides near Kerrville.
March 2 (1370); range "grassy places around St. Louis, Missouri; also in Arkansas and West Florida, Nuttall; Kentucky, Short."

Draba platycarpa T. & G. Fl. N. A. 1: 108 (1838).
Draba cuneifolia var. *platycarpa* S. Wats. Proc. Am. Acad. 23 : 256 (1888).
Collected with *D. cuneifolia* at Waco, the two often growing side by side, but readily distinguished. Found later at Corpus Christi along the railroad embankment of the S. A. & A. P.
March 8 (1411); type locality, "Texas."

ARABIS L. Sp. Pl. 664 (1753).

[TURRITIS L. Sp. Pl. 666 (1753).]
Arabis Virginica (L.) Trelease; Branner & Coville, Rep. Geol. Surv. Ark. 1884, 4: 165 (1891).
Cardamine Virginica L. Sp. Pl. 656 (1753).
Collected first at Waco, McLennan county, altitude 400 feet, on edges of cultivated fields (1372), and later at Corpus Christi, in open pasture land in moist soil, altitude 40 feet. Distributed as *Roripa* (*Nasturtium*) *tanacetifolium*.
March 8 (1407); type locality, "in Virginia."

SYNTHLIPSIS A. Gray, Mem. Am. Acad. 4: 116 (1849).

Synthlipsis Berlandieri hispida S. Wats. Proc. Am. Acad. 17 : 321 (1882).
Abundant in the low ground bordering on the flats at Corpus Christi, at sea level. Not observed on the plateau, where *Lesquerella Gordoni* takes its place.
March 8 (1405); from the type locality.

CAPPARIDACEAE.

POLANISIA Raf. Journ. Phys. 89: 98 (1819).

Polanisia trachysperma T. & G. Fl. N. A. 1: 669 (1840).
Low, sandy ground along the beach and along the streets at sea level, Corpus Christi, but not very abundant.
May 29 (1787); type locality, "Texas."

PLATANACEAE.

PLATANUS L. Sp. Pl. 999 (1753).

Platanus occidentalis L. Sp. Pl. 999 (1753).
Along the Guadalupe, at Kerrville, and other streams throughout Kerr county. Usually a tall slender tree.
April 19 (1622); type locality, "in America septentrionali."

ROSACEAE.

CRATAEGUS L. Sp. Pl. 475 (1753).

Crataegus Crus-galli L. Sp. Pl. 476 (1753).

On the summit of a stony, limestone hill about 5 miles northwest of Kerrville along Town Creek, altitude 1900 feet. Only two bushes about 5 feet high were seen. A western and southern extension of the range in Texas. "Extending into Texas to the Colorado and its tributaries."— Coulter.

April 29 (1668); type locality not given.

ROSA L. Sp. Pl. 491 (1753).

Rosa Arkansana Porter, Syn. Fl. Colo. 38 (1874).

Rosa blanda var. *Arkansana* Best, Bull. Torr. Club. **17** : 145 (1890).

In rich wooded bottom land along Bear Creek, northeast Kerr county, at an altitude of about 1800 feet. Two to three feet high, flowers white.

April 30 (1687); type locality, "banks of the Arkansas near Cañon City" (Colorado).

CERASUS Juss. Gen. 340 (1774).

Cerasus serotina (Ehrh.) Lois. in Duham. Nouv. **5** : 3 (1812).

Prunus serotina Ehrh. Beitr. **3** : 20 (1788).

Several small trees on the steep, left bank of the Guadalupe, at Kerrville, altitude 1625 feet.

April 19 in flower, May 16 in fruit (1592).

LEGUMINOSAE.

ACACIA Adans. Fam. Pl. 2 : 319 (1763).

Acacia filiculoides (Cav.) Trel.; Branner & Coville, Rept. Geol. Surv. Ark. 1888, **4** : 178 (1891).

Mimosa filiculoides Cav. Ic. **1**: 55, *t. 78* (1791).

Acacia filicina Willd. Sp. Pl. **4** : 1072 (1806).

Collected first at the Oso, in sandy ground at sea level (1563), where it was scarce, later at Kerrville, on the edges of cultivated fields, and under bushes in pasture land, altitude 1650–1800 feet.

May 21 (1770); type locality, "Mexico."

Acacia amentacea DC. Prodr. 2: 455 (1825).

Acacia rigidula Benth. Lond. Jour. Bot. 1: 504 (1842).

Usually a low, gnarled and twisted branching prostrate shrub at Corpus Christi, altitude 10-35 feet, beginning to flower before the leaves appear. Very spiny. Sometimes erect and spreading.

March 5 (1382); type locality, "in Nova-Hispania."— Mexico.

Acacia Farnesiana Willd. Sp. Pl. 4: 1083 (1806).

A handsome tree with smooth brown bark, much cultivated at Corpus Christi, but growing wild about the town, on the bluff portion, altitude 20-40 feet. The flowering heads on the specimens collected were only slightly odorous, rather small and lax, scattered.

March 17 (1454); type locality, "in Domingo."

Acacia Roemeriana Scheele, Linnaea 21: 456 (1848).

One of the most common shrubs on the hills about Kerrville, usually confined to near and on the summits, at an altitude of 1900-2000 feet. The original description calls for "flores rosei," perhaps owing to discoloration in the dried specimens. They are tawny white. Schlechtendahl instead of Scheele, is often given as the author of this species.

April 23 in flower, in fruit May 22 (1624); type locality, "prope Austin."

Acacia tortuosa Willd. Sp. Pl. 4: 1083 (1806).

Growing in company with, and flowering at the same time as *A. amentacea*. A prostrate, much-twisted, spreading bush, the orange yellow flowers of which are delightfully fragrant. The numerous heads are on slender, glandular peduncles an inch long. Spines over an inch long, whitish, at least the older ones. The pods are from two to four inches long, usually curved, linear, flat, with flat edges a line or two wide, tomentose, covered especially along the sides with cherry-colored glands. Flowering specimens of this have, no doubt, often been confused with *A. Farnesiana*. I took it to be that species until I found it in fruit.

March 5 (1383); type locality, West Indies.

MIMOSA L. Sp. Pl. 516 (1753).

Mimosa fragrans A. Gray, Bost. Jour. Nat. Hist. 6: 182 (1850).

Frequently met with along the stony banks of the Guadalupe and on hillsides at Kerrville, ranging from 1625-1900 feet altitude. A shrub 3-6 feet high, with slender, flexuous branches, bearing an abundance of pale rose purple flowers, which fade almost white when old. The pod is occasionally armed with scattered prickles.

April 19 in flower, May 20 in fruit (1594); type locality, "rocky soil on the Pierdenales," a stream eighteen miles north of Kerrville.

MORONGIA Britton, Mem. Torr. Club, **5**: 191 (1894).
[SCHRANKIA Willd. Sp. Pl. **4**: 1041 (1806), not Medic. (1792).]
Morongia angustata (T. & G.) Britton, Mem. Torr. Club, **5**: 191 (1894).
Schrankia angustata T. & G. Fl. N. A. **1** : 400 (1840).
Specimens referred to this species were collected in fruit at Kenedy, Carnes county, along the railroad embankment, altitude 400 feet, and at Corpus Christi in bare open ground along the beach at about 8 feet altitude.
May 26 (1779); type locality not given; range, S. Carolina; Georgia; Texas.
Morongia Roemeriana (Scheele).
Mimosa Roemeriana Scheele, Linnaea, **21** : 456 (1848).
Schrankia platycarpa A. Gray, Bost. Jour. Nat. Hist. **6** :183 (1850).
Abundant about Kerrville in stony or gravelly ground. *Mimosa Roemeriana* and *Schrankia platycarpa* are undoubtedly the same, as Gray suggests in his publication of *S. platycarpa*, but refused to recognize Scheele's plant because the fruit was not characterized. *Acacia Roemeriana* Scheele, described on the same page with *M. Roemeriana* is described from flowering material only, and in addition has the error of " flores rosei," yet it was accepted as a good species. The leaflets of the Kerrville plant are barely ciliate. They are like those of *M. angustata* in being veinless, but are shorter and broader. The unusually broad, flat legume, separates it at once from our other species.
April 23 in flower, June in fruit (1634); type locality, " prope New Braunfels."—Scheele; " dry, stony, prairies, New Braunfels."— Gray.

ACUAN Med. Theod. Sp. 52 (1786).
[DESMANTHUS Willd. Sp Pl. **4**: 1044 (1806).]
[DARLINGTONIA DC. Ann. Sci. Nat. **4** : 97 (1825).]
Acuan Illinoensis (Michx.) Kuntze, Rev. Gen. Pl. 158 (1891).
Mimosa Illinoensis Michx. Fl. Bor. Am. **2** : 254 (1803).
Acacia brachyloba Willd. Sp. Pl. **4** : 1071 (1806).
Desmanthus brachylobus Benth. in Hook. Journ. Bot. **4** : 358 (1842).
In low, moist ground along the Guadalupe at Kerrville, altitude 1600 feet. Flowers in lax heads, white or rose tinted.
June 13 (1847); type locality, "in pratensibus regionis Illinoensis."
Acuan depressa (H. & B.) Kuntze, Rev. Gen. Pl. 158 (1891).
Desmanthus depressus H. & B. in Willd. Sp. Pl. **4** : 1046 (1806).
Desmanthus diffusus Willd. Sp. Pl. **4** : 1046 (1806).
Plentiful in cultivated ground at the Oso, altitude about 15 feet.

Pros'rate and widely spreading. The majority of the pods are shorter than usual in my specimens.

April 12 (1553); type locality, "in America meridionali."

Acuan reticulata (Benth.) Kuntze, Rev. Gen. Pl. 158 (1891).

Desmanthus reticulatus Benth. in Hook, Journ. Bot. 4: 357 (1842).

Two plants of this species were collected along the roadside northeast of Kerrville, altitude 1700 feet. Two very good characters are omitted by Coulter, in the Manual of Western Texas, namely, "the glaucous color of the foliage in the fresh state," and the strongly reticulated pod, whence the specific name.

June 30 (1931); type locality, San Felipe, Texas, collected by Drummond.

Acuán velutina (Scheele) Kuntze, Rev. Gen. Pl. 158 (1891).

Desmanthus velutinus Scheele, Linnaea 21 : 455 (1848).

Common about Kerrville, from the low gravelly banks of Town Creek 1600 feet altitude, to about 1800 feet on the hillsides in pasture land. Usually decumbent; much branched from the stout perennial root.

April 23 to June 20 (1655); type locality, "auf sumpfigem Boden am oberen Komalkreek bei Neubraunfels."

NEPTUNIA Lour.

Neptunia lutea (Leavenw.) Benth. in Hook. Jour. Bot. 4: 356 (1842).

Acacia lutea Leavenw. Am. Jour. Sci. 7: 61 (1824).

Young plants were seen in sandy ground at Flower Bluff, Nueces county, in flower on April 9, at sea level. Later good fruiting specimens were obtained at Corpus Christi, growing in moist, rich black land on the edge of a water hole, near the Arroyo, altitude 40 feet.

June 5 (1822); type locality, "prairies of Green county, Alabama."

PROSOPIS L. Mant. 68 (1767).

Prosopis juliflora DC. Prodr. 2: 447 (1825).

Occurring at Corpus Christi in low ground outside of the "Bluff City Park" as a prostrate spreading bush flowering profusely. Also a good sized shrub near by, and on higher land often a small spreading tree. Very abundant throughout the region of Corpus Christi.

March 20 in flower, June 5 in fruit (1465); type locality, in siccioribus Jamaicae australis."

PARKINSONIA L. Sp. Pl. 375 (1753).

Parkinsonia aculeata L. Sp. Pl. 375 (1753).

This very ornamental tree was first noticed at Corpus Christi, beginning to flower early in April, altitude 35 feet. At San Antonio, where it is abundant, it was flowering profusely a month later.

April 11 to May 5 (1551); type locality, "in America calidiore."

CERCIS L. Sp. Pl. 374 (1753).

Cercis occidentalis Torr.; A. Gray. Bost. Jour. Nat. Hist. 6 : 177 (1850).

A large bush or small tree in rich ground about Kerrville, from the banks of Town Creek to the summits of the hills, altitude 1700 to 2000 feet.

April 26 (1653); from the type locality, "rocky plains of the Upper Guadalupe."

CASSIA L. Sp. Pl. 376 (1753).

Cassia Roemeriana Scheele, Linnaea 21 : 457 (1848).

This handsome species was first collected on the summits of hills about Kerrville; later it was found growing at much lower elevations, and quite plentiful. Always growing in rich open ground, altitude 1650–2000 feet.

April–June (1666); type locality, "auf felsigem Boden, am Rande von Gebüsch, nördlich von Neubraunfels."

CAESALPINIA L. Sp. Pl. 380 (1753).

[HOFMANSEGGIA Cav. Ic. 4 : 63, *t. 392–393* (1797).]

Caesalpinia Falcaria (Cav.) Fisher; Coult. Bot. Gaz. 18 : 122 (1893).

Hofmanseggia Falcaria Cav. Ic. 4 : 63, *t. 392* (1797).

Hofmanseggia stricta Benth.; A. Gray, Pl. Wright, 1 : 56 (1852).

A few plants were collected in a cultivated field at the Oso, in sandy ground, altitude about 6 feet.

April 9 (1554); type locality, "Zacatecas, Mexico."

BAPTISIA Vent. Dec. Gen. Nov. 9 (1808).

Baptisia australis (L.) R. Br. in Ait. f. Hort. Kew. 3 : 6 (1811).

Sophora australis L. Syst. Ed. 12, 2 : 287 (1767).

In low flat ground at the head of Nueces Bay, where it is abundant. Bears a dense drooping raceme of large creamy flowers, called "Meadow Queen" locally, and said to be injurious to stock if eaten. There was very little grass or other food where this was growing, yet it was untouched by the numerous cattle pasturing there.

April 3 (1523).

LUPINUS L. Sp. Pl. 721 (1753).

Lupinus subcarnosus Hook. Bot. Mag. *t. 3467* (   ).

Lupinus Texensis Hook. Bot. Mag. *t. 3492* (   ).

Along the road between Corpus Christi and the Oso, altitude 20 feet, growing on grassy banks. Also near the mouth of the Nueces, and on the hills about Kerrville. The roots bore tubercles in all the specimens examined.
March 21 (1466).

MEDICAGO L. Sp. Pl. 778 (1753).

Medicago sativa L. Sp. Pl. 778 (1753).
Occurring very sparingly about the streets of Kerrville, altitude 1650 feet.
June 13 (1845); type locality, "in Hispaniae."

MELILOTUS Juss. Gen. Pl. 356 (1789).

Melilotus Indica (L.) All. Fl. Ped. 1: 308 (1785).
Trifolium Melilotus Indica L. Sp. Pl. 765 (1753).
Along the banks of the San Antonio, at San Antonio, in rich, grassy, shaded ground. Apparently not plentiful.
May 5 (1701); type locality, in India.

PSORALEA L. Sp. Pl. 762 (1753).

Psoralea cuspidata Pursh, Fl. Am. Sept. 741 (1814).
Psoralea macrorhiza Nutt. in Fras. Cat. (1813), name only.
Psoralea cryptocarpa T. & G. Fl. N. A. 1: 301 (1838).
In stony limestone ground along the banks of the Guadalupe at Kerrville, on Town Creek, and on the plateau north of Kerrville, altitude 1625-1800 feet.
April 28 (1673); type locality, "in upper Louisiana."
Psoralea cyphocalyx A. Gray, Bost. Journ. Nat. Hist. 6: 172 (1850).
Plentiful, but scattered on the stony limestone hillsides about Kerrville, altitude 1700-1800 feet.
May 14 (1739); type locality, "Rocky prairies on the Cibolo and Pierdenales."
Psoralea esculenta, Pursh, Fl. Am. Sept. 475 (1814).
This plant occurs sparingly on the rocky hilltops, at an altitude of about 2000 feet. Found only in fruit. Low, 6-8 inches.
April 28 (1667); type locality, "on the banks of the Missouri."
Psoralea hypogea Nutt.; T. & G. Fl. N. A. 1: 302 (1838.).
Collected on a stony hillside north of Kerrville, altitude 1700 feet. Scarce.
April 28 (1677); type locality, "plains of the Platte."

Psoralea rhombifolia T. & G. Fl. N. A. 1 : 303 (1838).
In low, sandy ground at Corpus Christi and at the Oso, growing on the beach. Prostrate and widely spreading. Root ending in a deep-seated tuber. Flowers dark bronze.
April 9 (1462); type locality, Texas.

EYSENHARDTIA H. B. K. Nov. Gen. 6 : 489 (1823).

Eysenhardtia amorphoides H. B. K. Nov. Gen 6 : 491 (1823).
Plentiful in rich shaded ground at San Antonio along the river, altitude 600 feet; beginning to flower May 5. Also about Kerrville on the Guadalupe and on hillsides, altitude 1625-1900 feet.
June 12 (1705); type locality, "in Regno Mexicano, prope San Augustin de las Cuevas et Guanaxuato."

AMORPHA L. Sp. Pl. 713 (1753).

Amorpha fruticosa L. Sp. Pl. 713 (1753).
A bush 6-8 feet high, with slender, wand-like branches, on the banks of the Guadalupe in moist ground.
April 19 (1596); type locality, "in Carolina."

Amorpha subglabra (A. Gray).
Amorpha fruticosa var *subglabra* A. Gray, Bost. Jour. Nat. Hist. 6 : 174 (1850).
Amorpha Texana Buckley, Proc. Acad. Phila. 452 (1861).

A shrub 2-8 feet high branching above, young branches slightly pubescent, channeled, leaves almost horizontal, or slightly ascending, peduncles shorter than the lower pair of leaflets, pubescent with scattered spreading hairs; leaflets of the lower leaves often alternate, broadly oblong, ovate-oblong, or orbicular-ovate, on petiolules about one-eighth of their length, rounded or almost truncate at base, rounded or emarginate at apex, reticulated, smooth and shining above, the sparse pilose pubescence beneath especially noticable on the midvein, profusely punctate when fresh; flowering spikes to 2-4 inches long, dense, thick, calyx campanulate, glandular, pubescent, the teeth densely so with short white hairs; flowers rather large, twice the length of the calyx, deep purple, stamens much exserted; pod short, more than half enclosed in the calyx, obovate, rounded on the ventral side, almost straight on the dorsal.

A beautiful and well marked species, readily distinguished by the large coriaceous leaflets, and large and dense spike of dark purple flowers, fringed with the numerous exserted stamens, which bear reddish anthers. The leaves beneath are velvety to the touch.

Growing in rich ground 4 miles north of Kerrville, altitude 1800 feet, in copses along a dry water course. This station is about 20 miles south of the type station. The original description says "foliolis ellipticis retusis supra nitidis." Reference is also made to it in Pl. Wright. 1: 50 under *A. laevigata.* Buckley collected his type of *Amorpha Texana,* which is in the herbarium of the Academy of Natural Sciences at Philadelphia, and identical with my specimens, "near Dead Man's Hole, on the Pierdenales," a stream seven miles south of Fredericksburg.

May 21 (1772); type locality, "on a creek near Fredericksburg."

PAROSELA Cav. Desc. 185 (1803).

[DALEA Willd. Sp. Pl. 3: 1336 (1803), not P. Br. nor Gærtn.].

Parosela aurea (Nutt.) Britton, Mem. Torr. Club, 5: 196 (1894).

Dalea aurea Nutt. Fras. Cat. (1813).

Psoralea aurea Poir. in Lam. Encycl. Suppl. 4: 590 (1816).

Abundant in dry, stony limestone soil about Kerrville, altitude 1650–1900 feet.

June 14 (1856).

Parosela Hallii (A. Gray).

Dalea Hallii A. Gray, Proc. Am. Acad. 8: 625 (1873).

Found sparingly on dry stony slopes about Kerrville, altitude about 1800 feet. The plant is procumbent, from a very stout and long, branched, perennial root.

June 25 (1911); type locality, "on limestone, Dallas."

Parosela nana (Torr.).

Dalea nana Torr.; A. Gray, Pl. Fendl. 4: 31 (1849).

In sand at Flower Bluff near the mouth of Corpus Christi Bay, at sea level, and later a few specimens at Corpus Christi, in dry soil on the plateau near the Arroyo, altitude 40 feet.

April 9 (1535); type locality, "sandy soil, Willow Bar, on the Cimarron."

Parosela pogonathera (A. Gray).

Dalea pogonathera A. Gray, Pl. Fendl. 4: 31 (1849).

A few specimens of this species were picked up on a grassy bank at the Oso, at an altitude of about 8 feet.

March 21 (1482); type locality, "around Monterey, Mexico."

KUHNISTERA Lam. Encycl. 3: 370 (1789).

[PETALOSTEMON Michx. Fl. Bor. Am. 2: 48 (1803).]

Kuhnistera emarginata (T. & G.) Kuntze, Rev. Gen. Pl. 192 **i** (1891).

Petalostemon emarginatum T. & G. Fl. N. A. 1 : 311 (1838).

Found first in dry soil near the Arroyo at Corpus Christi, altitude 40 feet, and a few days later, growing in sand along the beach, almost at the water's edge. Much branched from the slender tap-root, which is apparently annual. Branches decumbent for a short distance, then ascending, giving a cup-shaped outline to the lower part of the plant.

May 31 (1799); type locality, "Texas."

Kuhnistera multiflora (Nutt.) A. A. Heller, Mem. Torr. Club, 5 : 197 (1894).

Petalostemon multiflorus Nutt. Journ. Phila. Acad. 7 : 92 (1834).

Along the beach of Corpus Christi Bay, on a grassy bank, growing in dense clumps. The root of this species is often very large and thick, sending out horizontal branches 1 to 2 feet long.

June 2 (1814); type locality, "in the plains of the Red river."

Kuhnistera pulcherrima n. n.
(PLATE 2.)

Petalostemon virgatum Scheele, Linnaea, 21 : 401 (1848), not Nees (1839).

Perennial, root stout in mature specimens, divergently branched, reddish brown; stems simple, erect, reddish, especially below, smooth, or sparingly pubescent; stipules filiform, subulate; leaves verticillastrate, smooth, the petioles about the length of the leaflets; petiolules very short; leaflets one to three pairs, usually a half inch in length, narrowly linear, slightly narrowed at each end, margins revolute in dried specimens, upper side dull green, glandular, especially along the margin; under side yellowish-green; peduncles rather short, two to four inches long; heads short-cylindrical, an inch or slightly more in length, as broad as long when in flower; bracts broadly ovate, shorter than the calyx, with brown, acuminate tips; calyx salmon or pinkish, with pubescent lines on each side, the lanceolate, or triangular-lanceolate, acute green lobes shorter than the tube, densely canescent; petals deep rose-purple; ovary slightly pubescent at base.

This plant has quite an interesting history. It is very plentiful about Kerrville, ranging from the banks of the Guadalupe to almost to the summits of the hills. I at once recognized it as something strange, and concluded that it was either new or *K. decumbens*, and as it did not prove to be that species, named it *K. pulcherrima* n. sp., and as such distributed it. The cause of this error is that Dr. Gray very much underrated the excellent work of Scheele, refusing, at least for a time, to recognize some very good species which he described. In this case he consigned Scheele's plant to synonomy, making it equal to *Petalostemon violaceus* var. *pubescens* A. Gray, a variety which was never published, and according to Gray's own statement, has no characters to separate it from the species, except more pubescence on the calyx. In one of the works, I think either Pl. Wright. or Pl. Fendl., I came across a reference to Scheele's species, which led me to think that my plant might be the same. Upon referring to the original description in Linnaea, it became evident that *Petalostemon virgatum*, which name, however, was used previously by Nees, is an excellent species, and agrees in all essential points with my specimens. New Braunfels, his type locality, is some sixty or seventy miles southeast of Kerrville, but the character of the country is the same. The plant can be distinguished from *K. purpurea* (*violaceus*) at a glance.

June 14 (1857).

INDIGOFERA L. Sp. Pl. 751 (1753).

Indigofera leptosepala Nutt.; T. & G. Fl. N. A. 1: 298 (1838).

Growing in sand along the beach near the Oso at sea level. The long, ashy-gray stems are procumbent and spreading. Seen also at San Antonio in open ground at 600 feet, and at Kerrville in stony, gravelly ground at 1600 feet.

April 9 (1544); type locality, "plains of Arkansas."

SESBANIA Pers.; Desv. Journ. Bot. 3: *t. 4* (1813–14).

Sesbania macrocarpa Muhl. Cat. 65 (1813).

Darwinia exaltata Raf. Fl. Lud. 106 (1817).

Found sparingly in wet ground on the right bank of the San Antonio, at the S. P. bridge. Young plants 1 to 2 feet high. The flowers are yellow in these specimens, instead of "yellow and red, dotted with purple."

June 9 (1831); type locality, "Car. Missis."

ASTRAGALUS L. Sp. Pl. 755 (1753).

Astragalus Brazoensis Buckley, Proc. Acad. Phila. 452 (1861).

Plentiful on a grassy bank at the "Blind Oso," and later at the Oso, in a cultivated field; altitude about 10 feet. The older specimens were long and procumbent, the early smaller ones erect.

March 21 (1483); type locality, "western Texas."

Astragalus leptocarpus T. & G. Fl. N. A. 1: 334 (1838).

In sandy soil near and along the beach of both Nueces and Corpus Christi Bays. Usually small, only five or six inches high.

March 12 (1443); type, locality, "near the Sabine river."

Astragalus Nuttallianus DC. Prodr. 2: 289 (1825).

Astragalus micranthus Nutt. Journ. Acad. Phila. 3: 122 (1821), not Desv.

Abundant in open ground at Corpus Christi; and always procumbent, not "ascending or erect"—Coulter. A poorly prepared specimen, not made with a view to showing the character of the plant, would make it appear as if erect.

March 5 (1380); type locality, "naked places in the prairies of Red river and the Arkansas."

Astragalus Wrightii A. Gray, Bost. Journ. Nat. Hist. 6: 176 (1850).

In dry stony ground along Town Creek and the Guadalupe, at Kerrville, altitude 1625 feet, growing near trees, but not directly shaded. Apparently a rare species.

April 23 (1633); type locality, "near Austin."

MEIBOMIA Adans. Fam. Pl. 2: 509 (1763).

[PLEUROBOLUS St. Hil. Bull. Soc. Philom. 1812, 192 (1812).]
[DESMODIUM Desv. Journ. Bot. 3: 122 (1813).]

Meibomia paniculata pubens (T. & G.) A. M. Vail, Bull. Torr. Club, 19: 112 (1892).

Desmodium paniculatum var. *pubens* T. & G. Fl. N. A. 1: 364 (1838).

Desmodium pubens Young, Fl. Texas 233 (1873).

On the left bank of the Guadalupe at Kerrville, in rich, open ground. Scarce.

June 20 (1893); type locality, "Tampa Bay, Florida."

Meibomia Tweedyi (Britton) A. M. Vail, Bull. Torr. Club. 19: 113 (1892).

Desmodium Tweedyi Britton, Trans. N. Y. Acad. Sc. 8: 183 (1890).

Specimens from the right bank of the Guadalupe, altitude 1600 feet, growing in gravelly ground, have been referred to this species. The flowers are yellow. Scarce.

June 13 (1846); type locality, Tom Green county.

LESPEDEZA Michx. Fl. Bor. Am. 2: 70 (1803).

Lespedeza repens (L.) Bart. Prodr. Fl. Phila. 2: 77 (1818).

Hedysarum repens L. Sp. Pl. 749 (1753).

Erect or reclining; rather plentiful in dry stony ground along Town Creek and on the bluff on the left bank of the Guadalupe, altitude 1625-1650 feet.
June 26 (1914).

VICIA L. Sp. Pl. 734 (1753).

Vicia Ludoviciana Nutt.; T. & G. Fl. N. A. 1: 271 (1838).
Common about Corpus Christi in sandy ground, from sea level to 40 feet.
March 27 (1513); type locality, "grassy places on the Red river."

RHYNCHOSIA Lour. Fl. Cochin. 562 (1793).

Rhynchosia minima (L.) DC. Prodr. 2: 385 (1825).
Dolichos minimus L. Sp. Pl. 726 (1753).
At San Antonio, along the S. P. R. R., near the bridge. Not given in Coulter's Manual of Western Texas. Growing in a tangled mass.
May 5 (1695); type locality, "in Jamaica."

Rhynchosia Texana T. & G. Fl. N. A. 1: 687 (1840).
About Kerrville, in stony ground along the banks of the Guadalupe, and in rich ground, edges of cultivated fields.
June 14 (1861); type locality, "Texas."

GERANIACEAE.

GERANIUM L. Sp. Pl. 676 (1753).

Geranium Carolinianum L. Sp. Pl. 682 (1753).
In an arroyo southeast of Corpus Christi, altitude about 15 feet, and about the streets of Kerrville. A low, spreading form.
March 27 (1510); type locality, "in Carolina, Virginia."

ERODIUM L'Hér. Geran. *t. 1* (1787).

Erodium cicutarium (L.) L'Hér.; Ait. Hort. Kew. 2: 414 (1789).
Geranium cicutarium L. Sp. Pl. 680 (1753).
In rich, shaded, low ground on the right bank of the Guadalupe at Kerrville. Seen only at this place, and scarce there.
May 14 (1743); type locality, "in Europae sterilibus cultis."

Erodium Texanum A. Gray, Bost. Journ. Nat. Hist. 6: 157 (1850).
In dry, open ground along Town Creek. Some of the plants were very large, procumbent, orbicular in outline, about a foot in diameter.
April 28 (1675); type locality, "small thickets in prairies above Victoria, and in patches in rocky soil at New Braunfels."

OXALIDACEAE.

OXALIS L. Sp. Pl. 433 (1753).

Oxalis dichondraefolia A. Gray, Pl. Wright. 1: 27 (1852).

Occasional around Corpus Christi, but growing in profusion in open ground at the "Blind Oso."

March 12 (1442); type locality, "Turkey Creek to the prairies of the San Felipe, and on the Rio Grande."

LINACEAE.

LINUM L. Sp. Pl. 277 (1753).

Linum multicaule Hook.; T. & G. Fl. N. A. 1; 678 (1840).

Linum selaginoides T. & G. Fl. N. A. 1: 205 (1838), not Lam.

Plentiful in dry open ground at Corpus Christi, altitude 10-40 feet. The earliest specimens are simple, one flowered, flowers rather large, dark orange in color.

March 5 (1389); type locality, "Texas."

Linum rigidum Pursh, Fl. Am. Sept. 210 (1814).

In low, sandy ground at Corpus Christi, sea level, and plentiful on the summits of ridges around Kerrville. Flowers orange color, lighter than those of *L. multicaule*. In Coulter's Manual of Western Texas, *L. rigidum* is placed under the section "** flowers rather small." The flowers are usually an inch in diameter as are those of the one placed next to it, *L. Berlandieri*.

April-June (1496); type locality, "on the Missouri."

Linum rupestre Engelm.; A. Gray, Bost. Jour. Nat. Hist. 6: 232 (1850).

On the steep stony left bank of the Guadalupe at Kerrville, altitude 1630 feet, profusely branching from a perennial root.

May 7 (1715); type locality, "growing from the crevices of naked rocks, New Braunfels."

MALPIGHIACEAE.

THRYALLIS L. Sp. Pl. Ed. 2, 554 (1762), not Ad. Juss.

[GALPHIMIA Cav. Ic. Desc. Pl. 5: 61 (1799).]

Thryallis angustifolia (Benth.) Kuntze, Rev. Gen. Pl. 89 (1891).

Galphimia angustifolia Benth. Bot. Sulph. 9, *t. 5* (1844).

Galphimia linifolia A. Gray, Bost. Journ. Nat. Hist. 6: 166 (1850).

On rocky ledges of the left bank of the Guadalupe, at Kerrville, and on stony, grassy hillsides, altitude 1640-1800 feet. This is the *G. linifolia* of Gray from the type locality, "rocky hills and prairies of the Upper Guadalupe."
May 14 (1737); type locality, "Cape San Lucas."

MALPIGHIA L. Sp. Pl. 425 (1753).

Malpighia glabra L. Sp. Pl. 425 (1753).
Found sparingly along Nueces Bay and in Corpus Christi, as a suffruticose plant 12-15 inches high, although good-sized bushes were noticed in cultivation.
March 5 (1396); type locality, in tropical America. "in Jamaica, Brasilia, Surinamo, Curacao."

ZYGOPHYLLACEAE.

PORLIERA Ruiz and Pav. Prodr. 55 *t. 9* (1794).

Porliera angustifolia (Engelm.) A. Gray, Pl. Wright. 1: 28 (1852).
Guiacum angustifolium Engelm. Wisliz. Rep. 29 (1848).
In shady soil near the mouth of the Nueces river, sea level to 30 feet. At a little distance this shrub or tree presents the appearance of a Conifer, its dark green leaves, and ascending branches giving it a very peculiar appearance.
April 3 (1524); type locality, "about Parras," Mexico.

TRIBULUS L. Sp. Pl. 386 (1753).

Tribulus maximus L. Sp. Pl. 386 (1753).
Tribulus terrestris Muhl. Cat. 43 (1818).
Tribulus trijugatus Nutt. Gen. 1: 277 (1818).
Kallstroemia maxima T. & G. Fl. N. A. 1: 213 (1838).
Plentiful in rich ground about Kerrville, altitude 1650 feet. Also abundant at Corpus Christi. Flowers open at about 10 o'clock on clear mornings and nearly an hour later when cloudy, remaining open about two hours.
May 23 (1777); type locality, "in Jamaicae aridis."

RUTACEAE.

THAMNOSMA Torr. & Frem. in Frem. 2d Rep. 313 ( ).

Thamnosma Texanum (A. Gray) Torr. Mex. Bound. Surv. 2: 42 (1859).

Rutosma Texana A. Gray, Gen. Fl. Am. Bor. Orien. 2 : 144 (1849).

In gravelly ground along Town Creek, 1600 feet and on the hilltops in rich rocky ground 2000 feet, but not observed at intermediate elevations. A homely little spreading plant, the greenish-yellow flowers apparently never fully expanding.

April 23 (1627).

ZANTHOXYLUM L. Sp. Pl. 270 (1753).

Zanthoxylum Fagara (L.) Sargent, Gard. and Forest, 3: 186 (1890).
Schinus Fagara L. Sp. Pl. 389 (1753).
Pterota subspinosa P. Br. Civ. & Nat. Hist. Jam. 146, *t. 5, f. 1* (1755).
Fagara Pterota L. Amoen. 5: 396 ( ).
Zanthoxylum Pterota H. B. K. Nov. Gen. 6: 3 (1823).

Collected near Gregory, San Patricio county, altitude 35 feet, where it is plentiful as a tall shrub or small tree. The leaves of these specimens are much smaller than are those from Florida, and the flower clusters much shorter. The bark is whitened, and in that particular at least, agrees with Buckley's *Z. hirsutum*, the type of which was collected in the same region, near Corpus Christi, but he does not state on which side of the Bay.

April 14 (1566); "in Jamaicae campestribus."

PTELEA L. Sp. Pl. 118 (1753).

Ptelea trifoliata L. Sp. Pl. 118 (1753).
In gravelly ground along Town Creek, altitude 1620 feet.
May 3 (1690); type locality, "in Virginia."
Ptelea trifoliata mollis T. & G. Fl. N. A. 1: 680 (1840).
In a copse at San Antonio, along the S. P. R. R., where it grows in abundance.
April 17 (1582); type locality, "Texas."

SIMARUBACEAE.

CASTELA Turp. Ann. Mus. Paris, 7 : 78, *t. 5* (1806).

Castela Nicholsoni Hook. Bot. Misc. 1: 271, *t. 56* (1830).
One of the chapparral bushes; plentiful about Corpus Christi, altitude 10–40 feet. Apparently not reported so far northeast before, as its range is given as "gravelly bluffs of the lower Rio Grande from Eagle Pass downwards."—Coulter.
March 8 (1402); type locality, West Indies.

MELIACEAE.

MELIA L. Sp. Pl. 384 (1753).

Melia Azederach L. Sp. Pl. 384 (1753).

This ornamental tree is cultivated extensively at Corpus Christi and at other places in Texas. A single tree was found growing along the roadside, between Corpus Christi and Nueces, altitude 30 feet. At Corpus Christi it begins to bloom about the middle of March, but at Kerrville, 270 miles northwest and 1600 feet higher, not until the latter part of April.

April 3 (1525); type locality, "in Syria."

POLYGALACEAE.

POLYGALA L. Sp. Pl. 701 (1753).

Polygala alba Nutt. Gen. 2: 87 (1818).

Very abundant in the stony limestone ground about Kerrville, ranging from the banks of Town Creek to almost the highest points on the hillsides, 1650–1900 feet.

April 24 (1645); type locality, "on the plains of the Missouri."

KRAMERIA Loefl. It. Hisp. 195 (1758).

Krameria secundiflora DC. Prodr. 1: 341 (1824).

Krameria lanceolata Torr. Ann. Lyc. N. Y. 2: 168 (1828).

Very plentiful, and having the same range as *Polygala alba*, except that it does not reach as great an altitude, its limit being about 1800 feet, stopping where the abrupt ascent of the ridges begins. The roots are usually thick, sending out very long branches which penetrate deep into the hard, stony ground. Flowers silky, maroon-colored.

April 23 (1625); type locality, "in Mexico."

EUPHORBIACEAE.

PHYLLANTHUS L. Sp. Pl. 981 (1753).

Phyllanthus polygonoides Spreng. Syst. 3: 23 (1826).

Plentiful in gravelly ground about Kerrville, at low and medium elevations, 1625–1700 feet.

April 25 (1640); type locality, Arkansas.

CROTON L. Sp. Pl. 1004 (1753).

Croton balsamiferus Willd. Sp. Pl. 4: 548 (1805).

Growing among chapparral at the "Blind Oso." Leaves somewhat viscid. A bush 3 or 4 feet high, at an altitude of 10-30 feet.

March 21 (1477); type locality, West Indies.

Croton capitatus Michx. Fl. Bor. Am. 2: 214 (1803).

Very abundant in dry ground near the Arroyo at Corpus Christi, in company with *Gomphrena Neallyi*.

May 30 (1800); type locality, "in regione Illinoensi."

Croton fruticulosus Engelm.; Torr. Mex. Bound Surv. 194 (1859).

In rich shaded ground on both banks of the Guadalupe at Kerrville. The leaves on these specimens are less pubescent than those of the type, probably due to the fact that the plants grew in moister and more shaded situations.

June 12 (1842); type locality, western Texas and northern Mexico.

Croton maritimus Walt. Fl. Car. 239 (1788).

In sand along the beach of Corpus Christi Bay, and the Gulf coast of Mustang Island, at sea level.

March—May (1423); type locality, Carolina.

Croton monanthogynus Michx. Fl. Bor. Am. 2: 215 (1803).

Very common at Kerrville along roadsides and in cultivated fields; in rich ground, altitude 1650-1700 feet. Seen also at Corpus Christi.

June 30 (1932); type locality, in Tennessee, near Nashville.

Croton punctatus Jacq. Coll. 1: 166 (1786).

Croton argyranthemus Michx. Fl. Bor. Am. 2: 215 (1803).

In sandy soil at Flower Bluff, altitude about 15 feet, growing among the scrub oaks.

April 9 (1547).

Croton suaveolens Torr. Mex. Bound. Surv. 2: 194 (1859).

At sea level along Nueces Bay, Nueces county, among chapparral. A branching bush about four feet high.

March 12 (1429); type locality, "on the Rio Grande."

Croton Texensis (Klotzch) Muell. Arg. in DC. Prodr. 15; Part 2, 692 (1862).

Hendecandra Texensis Klotzch in Erichs. Archiv. 1: 125 (1841).

Common in rich, open ground about Kerrville, altitude 1650 feet. Also seen at Corpus Christi and San Antonio.

June 15 (1863); type locality, Texas.

ACALYPHA L. Sp. Pl. 1003 (1753).

Acalypha Lindheimeri Muell. Arg. Linnaea 34: 47 (1865).
Acalypha phleoides Torr. Mex. Bound Surv. 2: 199 (1859), not Cav.
In stony and gravelly ground along the Guadalupe and Town Creek at Kerrville, altitude 1600 feet.
May 3 (1691); type locality, Texas.

Acalypha radians Torr. Mex. Bound Surv. 2: 200 (1859).
Common in bare exposed ground near the beach at Corpus Christi, where it grows prostrate from a tough, stout root. Along Nueces Bay, in rich ground where it was more protected, the stems were more slender, ascending.
April 3 (1519); type locality, "western Texas, especially along the Rio Grande."

TRAGIA L. Sp. Pl. 980 (1753).

Tragia ramosa Torr. Ann. Lyc. N. Y. 2: 245 (1824).
Tragia stylaris Muell. Arg. Linnaea, 34: 180 (1860).
Rather common in stony ground at Kerrville, along Town Creek and the Guadalupe, altitude 1600-1650 feet.
May 7 (1716); type locality, " sources of the Canadian ?"

JATROPHA L. Sp. Pl. 1006 (1753).

Jatropha spathulata (Ortega) Muell. Arg. in DC. Prodr. 15: Part 2, 1081 (1862).
Mozinna spathulata Ortega, Dec. 8: 105, *t. 13* (1797).
Very abundant at Corpus Christi; altitude, sea level to 40 feet. When cut, a wine colored juice exudes. Called "leather wood." Low, 1-2 feet high, stems soft and yielding.
April 11 (1550); type locality, Mexico.

STILLINGIA L. Mant. 1: 19 (1767).

Stillingia angustifolia (Muell. Arg.) Engelm.
In stony or gravelly ground about Kerrville, where it is plentiful, altitude 1600-1650 feet.
May 7 (1714).

Stillingia sylvatica L. Mant. 1: 126 (1767).
Growing in sand at Flower Bluff, and plentiful on Mustang Island on the low, western shore, at Rope's Pass. Leaves smaller and blunter than usual.
April 9 (1539); type locality, "in Carolinae pinetis."

ARGYTHAMNIA P. Br. Civ. & Nat. Hist. Jam. 338 (1755).
Argythamnia humilis (Engelm. & Gray) Muell. Arg. Linnaea, 34: 147 (1865).
   Aphora humilis Engelm. & Gray, Bost. Journ. Nat. Hist. 5: 262 (1845).
Collected at the Oso on a grassy bank, and in a cultivated field. Procumbent, older plants nearly two feet long.
March 21 (1484); type locality, "in hard clayey soil, west of the Brazos."
Argythamnia mercurialina (Nutt.) Muell. Arg. Linnaea, 34: 148 (1865).
   Aphora mercurialina Nutt. Trans. Am. Phil. Soc. (II.) 5: 174 (1837).
Scattered, in rich and usually shaded ground about Kerrville, altitude 1625–1800 feet.
April 24 (1648); type locality, "prairies of the Red river.

EUPHORBIA L. Sp. Pl. 450 (1753).
Euphorbia angusta Engelm.; Torr. Mex. Bound. Surv. 2: 189 (1859).
   Euphorbia Neallyi Coult. & Fisher, Cont. U. S. Nat. Herb. 2: 391 (1894).
In dry, stony ground about Kerrville, altitude 1625–1900 feet. Plentiful but scattered.
May 14 (1738).
Euphorbia campestris Cham. & Schlecht. Linnaea 5: 84 (1830).
   Euphorbia esulaeformis Schauer, Linnaea, 20: 729 (1847).
On the steep, stony left bank of the Guadalupe at Kerrville, altitude 1625 feet. The numerous erect stems from a stout, perennial root. Plentiful in this one situation, but not noticed elsewhere.
April 19 (1599); type locality, "in planitie inter Tlachichuca et Tepetitlan."
Euphorbia dentata Michx. Fl. Bor. Am. 2: 211 (1803).
Plentiful in rich ground at one time broken for a street, near Corpus Christi. Also abundant on the summits of hills about Kerrville. A much lower form than usual. The broader and shorter leaves nearly entire.
March 27 (1505); type locality, in Tennessee, near Nashville.
Euphorbia Fendleri T. & G. Pac. R. R. Rep. 2: 175 (1855).
Plentiful on a dry, stony slope along Town Creek, at Kerrville, altitude 1630 feet. The slender stems prostrate from a perennial root.
June 16 (1870).

Euphorbia maculata L. Sp. Pl. 455 (1753).

In low rich ground at Corpus Christi, altitude 40 feet, and at Kerrville in the gravelly bed of the Guadalupe, altitude 1600 feet.

May 30 (1804); type locality, North America.

Euphorbia nutans Lag. Gen. & Sp. 17 (1816).

Euphorbia Preslii Guss. Fl. Sic. Prodr. 1 : 539 (1827).

In the gravelly bed of the Guadalupe, at Kerrville, altitude 1600 feet. Ascending, usually stout and branching.

June 27 (1922).

Euphorbia obtusata Pursh, Fl. Am. Sept. 606 (1814).

In a grassy meadow at the Oso, altitude 10 feet. A rather common plant about Corpus Christi, in rich ground; often large and spreading.

March 21 (1475); type locality, Virginia, near Staunton.

Euphorbia polycarpa Benth. Bot. Sulph. 50 (1844).

Prostrate, growing in open ground at Corpus Christi, more often along the railroad embankment between the ties.

March 20 (1463); type locality, " Bay of Magdalena."

Euphorbia prostrata Ait. Hort. Kew. Ed. 1, 2 : 139 (   ).

Prostrate and spreading, in gravel, on the left bank of the Guadalupe, at Kerrville, altitude 1600 feet. My specimens agree well with several plants from Mexico, but do not altogether with the majority of the specimens noticed, which are from Florida and the eastern part of the country, and much smoother. Perhaps a distinct species.

June 27 (1918).

Euphorbia serpens H.B.K. Nov. Gen. 2 : 52 (1817).

On the plateau near the Oso, in rich ground, altitude 30 feet. Common about Corpus Christi, in bare, open ground.

March 21 (1467); type locality, "Cumanae prope Bordones et Punta Araya."

Euphorbia tetrapora Engelm. Mex. Bound. Surv. 2 : 191 (1859).

Plentiful in an arroyo southeast of Corpus Christi, growing with *E. obtusata*. Low and spreading, dark dull green in color.

March 27 (1509); range, Georgia to Texas.

ANACARDIACEAE.

RHUS L. Sp. Pl. 265 (1753).

Rhus aromatica Ait. Hort. Kew. 1 : 367 (1789).

Rhus Canadensis Marsh. Arb. Am. 129 (1785), not Mill.

A branching bush, occurring frequently on hillsides about Kerrville, altitude 1625-2000 feet.

April 26 (1658).

Rhus radicans L. Sp. Pl. 266 (1753).
 Rhus Toxicodendron radicans Marsh. Arb. Am. 131 (1785)
 On stony hillsides near Kerrville, altitude about 1800 feet. This was always a simple, erect shrub, about two feet high, with dull green, coriaceous leaves.
 April 28 (1658); type locality, "in Virginia, Canada."

AQUIFOLIACEAE.

ILEX L. Sp. Pl. 125 (1753).

Ilex decidua Walt. Fl. Car. 241 (1788).
 Plentiful through northeastern and central Texas, in low, damp ground. Collected without leaves, but bearing an abundance of orange-colored or red fruit, at Waco, McLennan county, altitude 400 feet, and on April 30 a few specimens in leaf were obtained along Bear Creek, Kerr county, altitude 1800 feet.
 March 2 (1374); type locality, Carolina.

HIPPOCASTANACEAE.

UNGNADIA Endl. Atakt. Bot. *t. 36* (1833).

Ungnadia speciosa Endl. Atakt. Bot. *t. 36* (1833).
 Along the stony, steep banks of the Guadalupe and Town Creek, near Kerrville. On nearly all the bushes observed, the flowers appeared before the leaves.
 April 19 (1598); type locality, in Texas.

SAPINDACEAE.

SAPINDUS L. Sp. Pl. 367 (1753).

Sapindus marginatus Willd. Enum. 432 (1809).
 Sapindus acuminatus Raf. New Fl. N. A. Part 3, 22 (1836).
 Sapindus falcatus Raf. Med. Bot. 2: 261 (1830).
 A slender, spreading tree about twenty feet high; plentiful along the Guadalupe about Kerrville.
 June 22 (1901).

CARDIOSPERMUM L. Sp. Pl. 366 (1753).

Cardiospermum Halicacabum L. Sp. Pl. 366 (1753).
 Creeping over bushes about Corpus Christi, where it is rather plentiful.
 June 2 (1817); type locality, "in Indiis."

RHAMNACEAE.

RHAMNUS L. Sp. Pl. 193 (1753).
Rhamnus Caroliniana Walt. Fl. Car. 101 (1788).
In moist, rich ground along the left bank of the Guadalupe at Kerrville, altitude 1600–1650 feet. A bush about six feet high.
May 14 (1740); type locality, Carolina.

CEANOTHUS L. Sp. Pl. 195 (1753).
Ceanothus ovatus Desf. Hist. Arb. 2: 381 (1809).
Ceanothus ovalis Bigel. Fl. Bost. Ed. 2, 92 (1824).
On the ledges of the left bank of the Guadalupe at Kerrville. A low, branching bush, about two feet high.
April 19 (1593).

COLUBRINA Rich.; Brongn. Ann. Sc. Nat. Ser. 1, 10: 368, *t. 15*, *f. 3* (1827).
Colubrina Texensis (T. & G.) A. Gray, Bost. Journ. Nat. Hist. 6: 169 (1850).
Rhamnus (?) *Texensis* T. & G. Fl. N. A. 1: 263 (1838).
Very plentiful about Corpus Christi, where it is usually a procumbent, twisted, spreading bush.
March in flower, June in fruit (1452); type locality, "Texas."

VITACEAE.

VITIS L. Sp. Pl. 202 (1753).
Vitis cordifolia Michx. Fl. Bor. Am. 2: 231 (1803).
Climbing high over bushes and trees along the Guadalupe at Kerrville.
May 15 (1750); type locality, not given. Range, from Pennsylvania to Florida.
Vitis monticola Buckley Proc. Acad. Phila. 450 (1861).
Abundant along hillsides and on summits about Kerrville. Shrubby, but climbing over low bushes. Apparently does not grow on low ground. Leaves less pubescent than in Buckley's type.
April 23 (1628); type locality, "mountainous districts of Burnet, Bell, and Hays counties."

AMPELOPSIS Michx. Fl. Bor. Am. 1: 159 (1803).
Ampelopsis arborea (L.) Rusby, Mem. Torr. Club, 5: 221 (1894).

Vitis arborea L. Sp. Pl. 203 (1753).
Ampelopsis bipinnata Michx. Fl. Bor. Am. 1: 160 (1803).
Cissus stans Pers. Syn. 1: 143 (1805).
Vitis bipinnata T. & G. Fl. N. A. 1: 243 (1838).
Climbing profusely over bushes at the Southern Pacific bridge at San Antonio.
June 9 (1829); type locality, "in Carolina, Virginia."
Ampelopsis cordata Michx. Fl. Bor Am. 1: 159 (1803).
Cissus Ampelopsis Pers. Syn. 1: 142 (1805).
Vitis indivisa Willd. Berl. Baumz. Ed. 2, 538 (1811).
Climbing over bushes on the edge of the Guadalupe at Kerrville.
June 12 (1841); type locality, Illinois.

MALVACEAE.

ABUTILON Gærtn. Fr. & Sem. 2: 251, *t. 135* (1791).

Abutilon Berlandieri A. Gray.

Stem rather tall and stout, suffruticose below. Scattered in low ground at Corpus Christi. The flowers open about five o'clock and close when it becomes dark.

June 5 (1824).

Abutilon incanum (Link) Sweet, Hort. Brit. 1: 53 (1826).
Sida incana Link, Enum. Hort. Berol. 2: 204 (1822).
Abutilon Texense T. & G. Fl. N. A. 1: 231 (1838).
Abutilon Nuttallii T. & G. Fl. N. A. 1: 231 (1838).

Apparently common throughout southern Texas. Usually rather slender, but stout, branching, two to four feet high. Flowers opening early in the afternoon and remaining open for several hours. Collected at San Antonio, altitude 600 feet.

June 9 (1837); type locality, Hawaiian Islands.

Abutilon Wrightii A. Gray.

Prostrate along the beach at Corpus Christi, near the upper end of the bay, altitude 10 feet. Flowers large, an inch or more in diameter, opening just before dark.

May 29 (1793).

CALLIRHOË Nutt. Journ. Acad. Phila. 2: 181 (1821).

Callirhoë digitata Nutt. Journ. Acad. Phila. 2: 181 (1821).

On stony limestone ridges and summits along Bear Creek, Kerr

county, altitude 1800 feet. Flowers white or pale lilac. Apparently scattered throughout Kerr county in similar situations.
April 30 (1685); type locality, "near Fort Smith."
Callirhoë involucrata (Nutt.); A. Gray, Mem. Am. Acad. (II.) 4: 15 (1848).
Nuttallia involucrata (Nutt.) Torr. Ann. Lyc. N. Y. 2: 172 (1825).
Malva involucrata T. & G. Fl. N. A. 1: 226 (1838).
Along the shore of Corpus Christi Bay at the Oso. Stems prostrate, 2–3 feet long, bearing a number of beautiful deep purple flowers. Found later along Nueces Bay, and in San Patricio county, growing in sandy ground near ant hills, altitude 35 feet.
March 21 (1468); type locality, "Valley of the Loup Fork."

MALVASTRUM A. Gray, Mem. Am. Acad. (II.) 4: 21 (1848).
[MALVEOPSIS Presl. Bot. Bem. 19 (1844)?]
Malvastrum Americanum (L.) Torr. Mex. Bound. Surv. 2: 38 (1859).
Malva Americana L. Sp. Pl. 687 (1753).
Malvastrum tricuspidatum A. Gray, Pl. Wright. 1: 16 (1852).
In rich ground at the Southern Pacific bridge, San Antonio, altitude 600 feet. Only a few plants were seen.
June 9 (1830); type locality, "in America."

SIDA L. Sp. Pl. 683 (1753).
Sida angustifolia Lam. Encycl. 1: 4 (1783).
Sida spinosa var. *angustifolia* Griseb. Fl. Brit. W. Ind. 74 (1859).
In dry ground at Corpus Christi, altitude 40 feet. This plant seemed scarce, as only a few were seen. Flowers copper-yellow, open during the forenoon. Stems ascending.
May 30 (1801); type locality, "Jamaica."
Sida ciliaris fasciculata (T. & G.) A. Gray, Proc. Am. Acad. 22: 294 (1887).
Sida fasciculata T. & G. Fl. N. A. 1: 231 (1838).
In dry sandy ground near Rockport, San Patricio county, altitude 35 feet, and at the Oso. Flowers open early in the forenoon, closing about eleven o'clock, dark lilac.
April 14 (1567); type locality, "Texas."
Sida diffusa H. B. K. Nov. Gen. 5: 257 (1821).
Sida filiformis Moric. Pl. Nouv. 38, *t.* 25 (1833–42), not Jacq.
Sida filicaulis T. & G. Fl. N. A. 1: 232 (1838).
A slender spreading procumbent plant, bearing small yellow flowers,

which are open only during the middle of the day. At Corpus Christi, in open ground near the beach, altitude about 10 feet. May 30 (1795), type locality; "prope Zelaya Mexicanorum."

Sida Helleri Rose, n. sp.

A low shrubby plant 3 cm. or less high, forming clumps 6 cm. in diameter; branches woody and procumbent, often covered with sand, and with erect, herbaceous, flowering shoots; leaves small, a little broader than long, 8 to 12 mm. long, rounded at apex, truncate or rounded at base, 3 to 5-nerved, coarsely crenate, more or less abundantly stellate-pubescent; petioles 6 to 12 mm. long; stipules persistent, foliaceous, linear, obtuse, 4 mm. long; flowers small, subsessile, solitary in the axis of the leaves; calyx campanulate, 5-lobed; sepals ovate, obtuse, 3 mm. long, in fruit 6 mm. long; corolla pale copper-colored, larger than the calyx; petals broad, somewhat oblique, glabrous; stamens united into a slender tube; styles 5, slender, with capitate stigmas; capsule deeply 5-lobed; carpels obtuse at tip, somewhat inflated, dehiscing at apex, one-seeded. Very common; flowers open about 4 P. M. Collected along the sandy shore of Corpus Christi Bay at the Oso, by A. A. Heller, April 9, 1894 (1533).

Very much like *S. cuneifolia* Gray (Pl. Wright. 7: 18), but with very different shaped leaves, much longer fruiting, calyx with obtuse instead of acute lobes, larger, more inflated and different shaped capsules.

The following note is from a letter of Mr. E. G. Baker, the well-known authority on Malvaceae:

"I have taken an opportunity of comparing your *Sida*, and I have very little to add from what you have already told me. It is a very interesting little plant, closely allied to *S. cuneifolia* A. Gray, but perfectly distinct. The shape, size and base of the leaves are different, and the calyx seems a good deal larger in your plant than in *S. cuneifolia*. It has the same ovate, membranous, slightly inflated carpels, so different from the *Eu Sideae*, and I suppose you will place it in the Section Pseudo-Malvastrum. I see there is generally one rather large leafy bract at the point of junction of the pedicel with the main stem."

J. N. Rose.

Sida physocalyx A. Gray, Bost. Jour. Nat. Hist. 6: 163 (1850).

In rich, open ground at Kerrville, altitude 1650 feet, about 75 miles south of the type locality. Stems numerous, and about 2 feet long in older plants, from a stout root. Flowers small, dull yellow, usually opening about eleven o'clock a. m., and remaining open an hour or less.

On a bright, warm day, the time of opening may be an hour or more earlier.

June 15 (1864); type locality, "on the Liano."

SPHAERALCEA St. Hil. Pl. Us. Bras. *t. 52* (1825).

Sphaeralcea Lindheimeri A. Gray, Bost. Jour. Nat. Hist. 6: 162 (1850).*

The first specimen observed was growing in sand on the beach, near the water's edge, at the Oso. Later in the day a few more plants were picked up in similar situations at Flower Bluff.

April 9 (1540).

MALVAVISCUS Dill.; Adans. Fam. Pl. 2: 399 (1763).

Malvaviscus Drummondii T. & G. Fl. N. A. 1: 230 (1838).

Pavonia Drummondii T. & G. Fl. N. A. 1: 682 (1840).

At San Antonio, along the river banks. This plant, at least at San Antonio, is not a shrub, as called for in the Manual of Western Texas. It is herbaceous, stout, about five feet high.

June 9 (1833); type locality, Texas.

CIENFUGOSIA Cav. Diss. 174, *t. 72, f. 2* (1787).

Cienfugosia sulphurea (St. Hil.) Garcke Bonpl. 8: 148 (1860).

Fugosia Drummondii A. Gray, Pl. Wright, 1: 23 (1852).

In rich, black land on the edge of a water hole near the Arroyo, Corpus Christi, altitude 40 feet. Very few plants were seen, and only one in flower, but the others in good fruit. Flower almost two inches in diameter, greenish-yellow. Apparently a very rare plant.

May 30 (1808); type locality, "Gonzales, Texas," for our plant.

VIOLACEAE.

CALCEOLARIA Loefl. Iter, 183–185 (1758).

[IONIDIUM Vent. Jard. Malm. *t. 27* (1803).]

Calceolaria verticillata (Ort.) Kuntze, Rev. Gen. Pl. 41 (1891).

---

* Type locality; "Victoria, on the lower Guadalupe." I do not know that this species has been collected since it was first described in 1850, until obtained by Mr. Heller. The type was collected by Lindheimer in 1845, but it had been previously obtained by Berlandier as early as 1834. The species had not been previously represented in the National Herbarium, and is probably the same with many of the herbariums. The type, which I have seen, is deposited in the Gray Herbarium along with 2 sheets from the Berlandier Herbarium. Dr. J. Gregg's No. 523 collected in 1848–9 seems to be a different species. Mr. Heller's plants were collected near Corpus Christi in April, 1894, and are in fine condition.   J. N. ROSE.

Viola verticillata Ort. Dec. Pl. 4: 50 (1797).
Ionidium polygalaefolium Vent. Jard. Malam. 27, *t.* 27 (1803).
Ionidium lineare Torr. Ann. Lyc. N. Y. 2: 168 (1827).
Plentiful about Corpus Christi from sea level to 40 feet. Usually procumbent and spreading.
March 9 (1414).

LOASACEAE.

MENTZELIA L. Sp. Pl. 516 (1753).

Mentzelia multiflora (Nutt.) A. Gray, Mem. Am. Acad. (II.) 4: 48 (1849).
Bartonia multiflora Nutt. Journ. Acad. Phila. (II). 1: 180 (1848).
On the left bank of the Guadalupe above Kerrville, in low stony ground, altitude 1600 feet. Plant stout, about 3 feet high.
June 21 (1896); type locality, "on the Rio Grande."

Mentzelia oligosperma Nutt.; Sims, Bot. Mag. *t. 1760* (1815).
Mentzelia aurea Nutt. Gen. 1: 300 (1818).
Low, dry ground, along Corpus Christi Bay. Stems long, weak, twining over bushes. Flowers bright copper-yellow, open during the middle of the day.
May 29 (1791); type locality, on the Missouri.

CACTACEAE.

ECHINOCACTUS Link & Otto, Verhand. Preiss. Gartenb. Verein, 3: 420 (1827).

Echinocactus setispinus hamatus Engelm. Proc. Am. Acad. 3: 272 (1856).
Growing on a sandy elevation at the Oso, in company with *Sida Helleri*. The yellow flowers open in the afternoon.
April 9 (1531).

Echinocactus Texensis Hoepf. in Allg. Gart. Zeit. 15: 297 (1842).
Echinocactus Lindheimeri Engelm.; A. Gray, Bost. Jour. Nat. Hist. 5: 246 (1845).
In sandy ground along Nueces Bay. Plants about eight inches in diameter. A handsome species with pink flowers.
April 3 (1532); type locality, Texas.

OPUNTIA Mill. Gard. Dict. Ed. 7 (1759).

Opuntia Engelmanni Salm-Dyck. Cact. Hort. Dyck 235 (1850).
Opuntia Lindheimeri Engelm.; A. Gray, Bost. Journ. Nat. Hist. 6: 207 (1850).

This most common of all Opuntias is plentiful throughout southern and central Texas. At some places between Waco and Kenedy, hundreds of plants can be seen from the car windows. About Corpus Christi the plants are usually large, and scattered in growth.

April 14 (1574): type locality, Mexico.

Opuntia Rafinesquii stenochila Engelm. Whipple's Exped. 43 (1856.)

In rich ground at Kerrville, altitude 1650-1700 feet. Flowers open widest during the middle of the day; pale yellow, with a reddish centre. Plants low.

May 15 (1749).

LYTHRACEAE.

AMMANNIA L. Sp. Pl. 119 (1753).

Ammannia auriculata Willd. Hort. Berol. 7, *t.* 7 (1806).

Ammannia Wrightii A. Gray, Pl. Wright. 2: 55 (1853).

In moist, rich black land, at Corpus Christi, on the edge of a water hole, altitude 40 feet. Plants small, 3-6 inches high.

June 5 (1821).

Ammannia coccinea Rottb. Pl. Hort. Havan. Descr. 7 (1773).

Ammannia latifolia T. & G. Fl. N. A. 1: 480 (1840), not L.

On the right bank of the Guadalupe at Kerrville, growing in mud and water.

June 28 (1925).

LYTHRUM L. Sp. Pl. 446 (1753).

Lythrum alatam Pursh, Fl. Am. Sept. 334 (1814).

About Corpus Christi, usually in moist ground, altitude 40 feet, and at Kerrville along the Guadalupe, altitude 1600 feet, where it was much taller, more erect and slender. Distributed as *L. lanceolatum* Ell.

March 27 (1506); type locality, "in lower Georgia."

Lythrum ovalifolium Engelm.; A. Gray, Bost. Jour. Nat. Hist. 6: 187 (1850).

Lythrum alatum var. *ovalifolium* A. Gray, Bost. Jour. Nat. Hist. 6: 187 (1850).

On the right bank of the Guadalupe at Kerrville, growing in low, wet ground, altitude 1600 feet.

June 19 (1885); type locality, "springs of the Pierdenales on rocks covered by water."

ONAGRACEAE.

ŒNOTHERA L. Sp. Pl. 346 (1753).

Œnothera Drummondii Hook. Bot. Mag. *t. 3361* (   ).
Growing in sand on the beach at Corpus Christi. Stems always procumbent, flowers about two inches in diameter, opening just before dark. Called "Buttercups."
March 27 (1512); type locality, Texas.

XYLOPLEURUM Spach, Hist. Veg. 4: 378 (1835).

Xylopleurum roseum (Ait.) Raimann, in Engler & Prantl. Nat. Pfl. Fam. 3: Abt. 7, 214 (1893).
Œnothera rosea Ait. Hort. Kew. 2: 3 (1789).
On the banks of the San Antonio, growing in grass. Found only in fruit.
May 5, (1703).

MEGAPTERIUM Spach, Hist. Veg. 4: 363 (1835).

Megapterium Missouriensis (Sims) Spach, Hist. Veg. 4: 364 (1835).
Œnothera Missouriensis Sims, Bot. Mag. *t. 1592* (1814).
Œnothera macrocarpa Pursh, Fl. Am. Sept. 734 (1814).
On rocky hillsides and lower summits about Kerrville. A night bloomer, but the large yellow flower, three inches in diameter, can still be found early the next morning.
April 23 (1629).

MERIOLIX Raf. Am. Month. Mag. 4: 192 (1818).

[CALYLOPHUS Spach, Hist. Veg. 4: 349 (1835).]

Meriolix serrulata (Nutt.) Walp. Rep. 2: 79 (1843).
Œnothera serrulala Nutt. Gen. 1: 246 (1818).
Calylophus Nuttallii Spach, Hist. Veg. 4: 350 (1835).
At Corpus Christi in low dry ground along the beach, but not plentiful.
April 2 (1517); type locality, "from the river Platte to the mountains."

Meriolix spinulosa (Nutt.).
Œnothera spinulosa Nutt.; T. & G. Fl. N. 1: 502 (1840).
Meriolix serrulata var. *spinulosa* Walp. Rep. 2: 79 (1843).
Œnothera serrulata var. *pinifolia* Engelm.; A. Gray, Bost. Jour. Nat. Hist. 6: 189 (1850).

Plentiful about Kerrville, especially along the Guadalupe and Town Creek, usually in gravelly or stony ground. Not only the throat of the calyx and the disk-shaped stigma are dark black-purple, but also the throat of the corolla. Of the hundreds of flowers seen, hardly a half dozen were without this marking. The variety *pinifolia* is merely a very narrow leaved form of this species. Both forms grow together and there is no other character to distinguish them.

April 19 (1600); type locality, Arkansas.

GAURA L. Sp. Pl. 347 (1753).

Gaura Drummondii (Spach) T. & G. Fl. N. A. 1: 519 (1840).
Schizocarya Drummondii Spach, Monog. Onogr. 62 ( ).
Gaura Roemeriana Scheele, Linnaea, 21: 579 (1848), *fide* Watson's Index.

In grassy woodland at San Antonio, along the S. P. R. R., altitude 600 feet. Rather plentiful.

April 17 (1590); type locality, Texas.

Gaura parviflora Dougl.; Hook. Fl. Bor. Am. 1: 208 (1833).
Along the left bank of the Guadalupe, at Kerrville, altitude 1600 feet, but not plentiful.

May 19 (1768); type locality, "sandy banks of the Wallawallah river."

Gaura sinuata Nutt.; Ser. in DC. Prodr. 3: 44 (1828).
About Kerrville in rich ground, altitude 1650–1750 feet. Plentiful.

May 3 (1692); type locality, Arkansas on the Red river.

Gaura suffulta Engelm.; A. Gray, Bost. Jour. Nat. Hist. 6: 191 (1850).
Plentiful at Corpus Christi, in dry, low ground at sea level. Flowers rather small, but the fruit agrees with specimens of *G. suffulta*.

March 10 (1391); type locality, New Braunfels.

UMBELLIFERAE.

DAUCUS L. Sp. Pl. 242 (1753).

Daucus pusillus Michx. Fl. Bor. Am. 1: 164 1803).
Plentiful in low sandy ground about Corpus Christi, especially in cultivated fields.

April 12 (1438); type locality, "in campestribus Carolinae."

BIFORA Hoffm. Umb. Gen. Ed. 2, 191 (1816).

Bifora Americana (DC.) Benth. & Hook. f.; S. Wats. Bibl. Index, 415 (1878).

Atrema Americana DC. Mem. Omb. 71, *t. 18* (1829).
Abundant on hilltops about Kerrville, in rich, stony ground.
May 21 (1656).

POLYTÆNIA DC. Mem. Omb. 53, *t. 13* (1829).
Polytaenia Nuttallii DC. Mem. Omb. 54, *t. 13* (1829).
Along the Guadalupe, and on rocky slopes and summits about Kerrville, altitude 1600-1900 feet.
April-June (1669).

CYNOSCIADUM DC. Mem. Omb. 44, *t. 11* (1829).
Cynosciadum pinnatum pumilum Engelm. & Gray, Bost. Jour. Nat. Hist. 5: 218 (1845).
In moist open ground at Corpus Christi. A small, procumbent plant.
March 10 (1409); type locality, "prairies, Galveston."

SANICULA L. Sp. Pl. 235 (1753).
Sanicula Canadensis L. Sp. Pl. 235 (1753).
In rich shaded ground along the river at San Antonio. Distributed as *S. Marylandica*.
May 5 (1713); type locality, in Virginia.

AMMOSELINUM T. & G. Pac. R. R. Rept. 2: 165 (1855).
Ammoselinum Popei T. & G. Pac. R. R. Rept. 2: 165 (1855).
Apium Popei A. Gray, Proc. Am. Acad. 7: 343 (1868).
In grassy, sandy ground along Corpus Christi Bay, at sea level.
March 21 (1474); type locality, "Llano Estacado."

CHAEROPHYLLUM L. Sp. Pl. 258 (1753).
Chaerophyllum procumbens dasycarpum (Nutt.) Coult. & Rose, Bot. Gaz. 12: 160 (1887).
    Chaerophyllum dasycarpum Nutt.; T. & G. Fl. N. A. 1: 638 (1842).
    Chaerophyllum Tainturieri var. [*dasycarpum*] Hook.; S. Wats. Bibl. Index 416 (1878).
In rich ground on a bank along Nueces Bay, where it was plentiful; altitude 15 feet. Also at San Antonio in rich, shaded ground near the S. P. bridge. This erect, rather stout plant, is very different in habit from our weak and procumbent *C. procumbens*.
March 12 (1521).

APIUM L. Sp. Pl. 264 (1753).

Apium leptophyllum (DC.) F. Muell.; Benth. Fl. Austral. 3: 372 (1866).

Sison Ammi L. Sp. Pl. 252 (1753). ?
Heliosciadum leptophyllum DC. Prodr. 4: 105 (1830).

In low, dry, grassy ground at Corpus Christi and the Oso; sea level to 20 feet.

April 12 (1560); type locality, North America.

SPERMOLEPIS Raf. Neog. 2 (1825).

[LEPTOCAULIS Nutt.; DC. Mem. Omb. 39, *t. 10* (1829).]

Spermolepis divaricatus (Walt.) Britton, Mem. Torr. Club, 5: 244 (1894).

Daucus divaricatus Walt. Fl. Car. 114 (1788).
Leptocaulis divaricatus DC. Mem. Omb. 39, *t. 10* (1829).
Apium divaricatum Wood, Bot. & Flor. 140 (1870).
Leptocaulis diffusus Nutt.; DC. Prodr. 4: 107 (1830).

Plentiful about Kerrville on stony hilltops in rich ground.

May 21 (1773); type locality, on the Red River.

Spermolepis echinatus (Nutt.).

Leptocaulis echinatus Nutt.; DC. Prodr. 4: 107 (1830).

At the Oso in grassy, sandy ground. Plentiful there, and also seen about Corpus Christi.

April 12 (1561); type locality, on the Red River.

CARUM L. Sp. Pl. 263 (1753).

Carum Petroselinum Benth. & Hook. Gen. Pl. 1: 890 (1867).

At San Antonio in rich ground along the river bank. Escaped from cultivation.

June 9 (1838).

PTILIMNIUM Raf. Jour. Phys. 89: 258 (1819).

[DISCOPLEURA DC. Mem. Omb. 38 (1829).]

Ptilimnium laciniatum (Engelm. & Gray) Kuntze, Rev. Gen. Pl. 269 (1891).

Daucosma laciniatum Engelm. & Gray, Bost. Jour. Nat. Hist. 6: 211 (1850).

On stony ridges around Kerrville, especially just below the summits, also a few plants seen along the river bank. Altitude 1600–1900 feet.

July 4 (1943); type locality, "near New Braunfels."

BOWLESIA Ruiz & Pav. Prod. Pl. Per. 44, *t. 34* (1794).

Bowlesia lobata Ruiz and Pav. Fl. Peruv. 3: 28, *t. 251* (1802).
Growing in rich ground under trees and bushes at Corpus Christi.
March 23 (1493); type locality, Peru.

HYDROCOTYLE L. Sp. Pl. 234 (1753).

Hydrocotyle prolifera Kellogg, Proc. Cal. Acad. 1: 14 (1873).
Hydrocotyle vulgaris Cham. & Schlecht. Linnaea, 1: 356 (1826), not L.
Hydrocotyle interrupta T. & G. Fl. N. A. 1: 599 (1840), in part.
In wet ground along the Guadalupe, at Kerrville, altitude 1600 feet.
July 2 (1935).

CORNACEAE.

CORNUS L. Sp. Pl. 117 (1753).

Cornus asperifolia Michx. Fl. Bor. Am. 1: 93 (1803).
Cornus Drummondii C. A. Meyer, Mem. Acad. St. Petersb. (VI.) 5: 210 (1845).
On the left bank of the Guadalupe, in rich moist ground. A spreading bush, 6–10 feet high.
May 7 (1717); type locality, South Carolina.

PRIMULACEAE.

SAMOLUS L. Sp. Pl. 171 (1753).

Samolus alyssoides n. sp.

(PLATE 3.)

Low, 6–8 inches high; purplish, especially the lower part of the stem and petioles; smooth and glaucous, branching from the base, erect; stems very leafy below; leaves crowded, more or less verticillate, spatulate-obovate, usually acutish, tapering into a broad, margined petiole, clasping at base, thick and coriaceous, the width at the widest part about one-third of the length; calyx slightly longer than the ovary, the triangular-lanceolate, acute lobes equalling the tube; flower small, white, like our other members of the genus; a cluster of glands at the base of each petal lobe; stigma entire, slightly thickened.

Related to *S. ebracteata*, the shape of the leaves and their manner of growth being much the same, but they differ in being more clustered at the base of the stems. It too is destitute of sterile filaments, but is stouter in every way, lower, more erect, with shorter, slightly thicker, ascending pedicels, larger flowers and capsules, and entire stigma. Unlike *S. ebracteata*, it grows only in dry, open, exposed ground, and only near salt water.

Collected along the beach at the upper end of Corpus Christi Bay, where it is scattered. Extremely plentiful on the low, sandy west shore of Mustang Island at Rope's Pass, growing in clumps.

May 29 (1788).

Samolus ebracteatus H.B.K. Nov. Gen. 2: 223, *t. 129* (1817).

On wet limestone rocks on the left bank of the Guadalupe at Kerrville, altitude 1600 feet. Stems weak and fleshy, long and slender, more or less leafy, reclining.

May 16 (1751); type locality, southern shores of Cuba.

Samolus floribundus H.B.K. Nov. Gen. 2: 224 (1817).

Samolus Valerandi var. *Americanus* A. Gray, Man. Ed. 2, 274 (1856).

On the banks of the Guadalupe in mud, at the water's edge; not abundant.

June 12 (1843); type locality, in Peru, near Callao and Lima.

SAPOTACEAE.

BUMELIA Sw. Prodr. 49 (1788).

Bumelia monticola Buckley, Bull. Torr. Club, 10: 91 (1883).

On the left bank of the Guadalupe, on moist, rocky soil. A spreading bush five to six feet high. The leaves are slightly pubescent on the veins, otherwise it is like the type.

June 2 (1938); type locality, " Mountains of El Paso county."

EBENACEAE.

DIOSPYROS L. Sp. Pl. 1057 (1753).

Diospyros Texana Scheele, Linnæa, 22: 145 (1849).

In low, dry ground, at Corpus Christi, as a spreading gnarled bush, 2 feet high; at the head of Nueces Bay a slender bush, 8 feet high; near Gregory, San Patricio county, and at San Antonio and Kerrville as small trees, 10–15 feet high. Altitude, sea level to 1600 feet.

March 12 (1431); type locality, New Braunfels, Texas.

OLEACEAE.

FRAXINUS L. Sp. Pl. 1057 (1753).
Fraxinus lanceolata Borck. Handb. Forst. Bot. 1: 826 (1800).
Fraxinus viridis Michx. f. Hist. Arb. 3: 115, *t. 10* (1813).
Fraxinus juglandifolia Willd. Sp. Pl. 4: 1104 (1806), not Lam.
A small tree, growing on the river bank at San Antonio, altitude 600 feet.
May 5 (1711).

LOGANIACEAE.

MENODORA Humb. & Bonpl. Pl. Æquin. 2: 98, *t. 110* (1809).
Menodora heterophylla Moric.; DC. Prodr. 8: 316 (1844).
Bolivaria Grisebachii Scheele, Linnaea, 25: 254 (1852).
In dry ground at Corpus Christi, from sea level to 40 feet, usually growing in patches. Flowers opening in the morning.
March 5 (1390); type locality, Mexico.
Menodora longiflora A. Gray, Amer. Jour. Sci. (II.) 14: 43 (1852).
On the steep, stony left bank of the Guadalupe at Kerrville, altitude 1625 feet. Usually growing in clumps; flowers opening late in the afternoon, often remaining open until the middle of the next forenoon.
June 18 (1880); type locality, "Texas."

SPIGELIA L. Sp. Pl. 149 (1753).
Spigelia Texana (T. & G.) A. DC. Prodr. 9: 5 (1845).
Coelostylis Texana T. & G. Fl. N. A. 2: 44 (1842).
Under a bush along the road about 4 miles northeast of Kerrville. Seen also near Corpus Christi.
May 8 (1719); type locality, "Texas."

POLYPREMUM L. Sp. Pl. 111 (1753).
Polypremum procumbens L. Sp. Pl. 111 (1753).
In rich, dry ground on the edge of a water hole at Corpus Christi, altitude 40 feet.
May 30 (1805); type locality, "in Carolina, Virginia."

GENTIANACEAE.

ERYTHRAEA Neck. Elem. 2: 10 (1790).
Erythraea Beyrichii T. & G.; Torr. in Marcy's Rep. 291, *t. 13* (1853).

Erythraea tricantha var. *angustifolia* Griseb.; DC. Prodr. 9: 60 (1845).
Hanging from wet limestone rocks on the left bank of the Guadalupe at Kerrville. Radical leaves in rosulate tufts.
July 2 (1940); type locality, "on the Washita."

Erythraea calycosa nana A. Gray, Syn. Fl. 2: 113 (1878).
In rich, stony limestone soil, summits of hills about Kerrville, altitude 2000 feet. Plentiful, growing in patches.
June 18 (1876); type locality, "W. Texas."

EUSTOMA Salisb. Parad. Lond. *t. 34* (1806).
Eustoma exaltata (L.) Griseb. DC. Prodr. 9: 51 (1845).
Gentiana exaltata L. Sp. Pl. Ed. 2, 331 (1762).
Lisianthus exaltatus Lam. Ill. 1: 478 (1791).
Eustoma silenifolium Salisb. Parad. Lond. *t. 34* (1806).
In rich, grassy ground on the right bank of the San Antonio river, at the Southern Pacific bridge at San Antonio. The large flowers are pale mauve in color. Scarce.
June 9 (1834).

ASCLEPIADACEAE.

ASCLEPIAS L. Sp. Pl. 214 (1753).
Asclepias longicornu Benth. Pl. Hartw. 24 (1840).
Near Corpus Christi, especially along the railroad. A low plant, decumbent at base, from a thick and fleshy tuberous root.
April 14 (1575); type locality, Mexico.

Asclepias Texana n. sp.

(PLATE 4.)

Perennial, the main root sending out fibrous rootlets; stem erect, slightly woody at base, two to three feet high, cymosely branched above, glaucous, purplish below, green above and marked with one or two pubescent lines, otherwise glabrous; leaves opposite, oval or ovate, acute or the lowest obtusish, and broader, somewhat oblique at base; petioles about one-fifth the length of the blade; peduncles comparatively stout, shorter than the leaves; umbels 15-20 flowered, on pedicels nearly half the length of the peduncles; flowers white with prominently exserted horns.

A beautiful species related to the northern *A. quadrifolia.* At first it was thought that it might be the West Indian *A. nivea* L., but reference to the plate on which that species was founded, to specimens in the Herbarium of Columbia College, and to descriptions, show that it is not that plant. *A. perennis* is perhaps its nearest neighbor in some respects, but has smaller flowers, and leaves tapering at both ends; besides, its range is eastern, and it grows in low ground.

In the Herbarium of Columbia College is a plant from either western Texas or New Mexico, referable to this species, although the flowers are smaller. Dr. Gray named it *A. perennis.* In the U. S. National Herbarium is a specimen of *A. Texana* collected by Lindheimer near New Braunfels, but unnamed if I remember rightly.

Collected on limestone hillsides about Kerrville, in ground shaded by trees and bushes, and along Town Creek in similar situations, altitude 1600–1800 feet.

June 14 (1859).

ASCLEPIODORA A. Gray, Proc. Am. Acad. 12: 66 (1876).

Asclepiodora decumbens (Nutt.) A. Gray, Proc. Am. Acad. 12: 66 (1876).

Anantherix decumbens Nutt. Trans. Am. Phil. Soc. (II.) 5: 202 (1833–37).

Rather abundant in dry, stony soil about Kerrville.

April 23 (1631); type locality, "near the confluence of the Kiamesha and Red rivers."

Asclepiodora viridis (Walt.) A. Gray, Proc. Am. Acad. 12: 66 (1876).

Asclepias viridis Walt. Fl. Car. 107 (1788).

Occasional in rich ground near Kerrville, especially in wooded pasture land. Some forms had narrow leaves approaching those of *A. decumbens.*

May 8 (1722); type locality, Carolina.

ACERATES Ell. Bot. S. C. & Ga. 1: 316 (1817).

Acerates viridiflora (Raf.) Eaton, Man. Ed. 5, 90 (1829).

Asclepias viridiflora Raf. Med. Rep. (II.) 5: 360 (1808).

Scarce, along the steep, stony left bank of the Guadalupe at Kerrville, altitude 1630 feet.

June 26 (1913).

Acerates angustifolia (Nutt.) Dec. in DC. Prodr. 8: 522 (1844).
Polyotus angustifolius Nutt. Trans. Am. Phil. Soc. (II.) 5: 201 (1833-37).
Acerates auriculata Engelm.; Torr. Mex. Bound. Surv. 2: 160 (1859).
Asclepias stenophylla A. Gray, Proc. Am. Acad. 12: 72 (1876).
In dry, gravelly or stony ground along Town Creek and the Guadalupe at Kerrville, but scarce.
June 16 (1868).

AMPELANUS Raf.; Britton, Bull. Torr. Club, 21: 314 (1894).
[ENSLENIA Nutt. Gen. 1: 164 (1818), not Raf.]
Ampelanus ligulatus (Benth.).
Enslenia ligulata Benth. Pl. Hartw. 290 (1848).
In rich ground on the left bank of the Guadalupe above Kerrville, twining over bushes and low trees. Apparently not previously reported within the borders of the United States.
June 21 (1899); type locality, "ad Aguas Calientes."

METASTELMA R. Br. Mem. Wern. Soc. 1: 52 (1809).
Metastelma barbigerum Scheele, Linnaea, 21: 760 (1848).
At the Oso, climing over bushes, altitude 10 feet. Not plentiful.
April 12 (1559); type locality, New Braunfels, Texas.

VINCETOXICUM Walt. Fl. Car. 104 (1788).
[GONOLOBUS Michx. Fl. Bor. Am. 1: 119 (1803).]
Vincetoxicum biflorum (Raf.).
Gonolobus biflora Raf. New Fl. 4: 58 (1836).
Chthamalia biflora Dec.; DC. Prod. 8: 605 (1844).
In rich limestone ground along Bear Creek, Kerr county, altitude 1800 feet. *Gonolobus biflora* is credited to Nuttall, in Decaisne, DC. Prod. as "Nutt.? in DC. herb.," and by Gray is cited as published in Torrey, Mex. Bound. Surv. 2: 165 (1859), but Rafinesque is totally ignored.
April 30 (1681); type locality, Red River, Arkansas and Texas.
Vincetoxicum reticulatum (Engelm.).
Gonolobus reticulatus Engelm.; A. Gray, Proc. Am. Acad. 12: 75 (1876).
Gonolobus granulatus Torr. Mex. Bound. Surv. 165 (1859), not Scheele.

Rather plentiful in rich, shaded ground about Kerrville, climbing over bushes and small trees.

April 24 (1644); type locality, "mountain ravine near Live Oak"

CONVOLVULACEAE.

IPOMOEA L. Sp. Pl. 160 (1753).

Ipomoea Lindheimeri A. Gray. Syn. Fl. 2: 210 (1878).

Ipomoea heterophylla Torr. Mex. Bound. Surv. 2: 149 (1859), not Ortega.

On stony limestone ridges northeast of Kerrville, altitude 1900 feet. Twining over bushes. Corolla light blue, but turning pink when dry.

May 21 (1776); type locality, west Texas.

CONVOLVULUS L. Sp. Pl. 153 (1753).

Convolvulus incanus Vahl. Symb. Bot. 3: 23 (1794).

In dry, open ground about Kerrville, altitude 1650-1750 feet.

June 23 (1910); range, Arkansas to Texas and Arizona.

EVOLVULUS L. Sp. Pl. Ed. 2, 391 (1762).

Evolvulus Nuttallianus R. & S. Syst. 6: 198 (1820).

Evolvulus pilosus Nutt. Gen. 1: 174 (1818), not Roxb.

Evolvulus argenteus Pursh, Fl. Am. Sept. 187 (1814), not R. Br.

In dry, bare places about Kerrville, altitude 1650-1800 feet. Corolla pale purple.

June 23 (1912); type locality, "confluence of the Rapid river and the Missouri."

Evolvulus sericeus Swartz, Prodr. Fl. Ind. Occ. 55 (1783-87).

In sandy ground at sea level about Corpus Christi. Flowers white.

March 12 (1441); type locality, Jamaica.

CRESSA L. Sp. Pl. 223 (1753).

(PLATE 5.)

Cressa aphylla n. sp.

From an apparently perennial root; low, about six inches high, slender, diffusely branched from the base; whole plant covered with scales and appressed hairs; leafless, each branch subtended by an ovate, acute or acutish bract or scale, smaller ones scattered along the naked branches and at the base of each flowering pedicel; flowers on very short pedicels; calyx bibracteolate, the bracts appressed, lanceolate or ovate lanceolate, barely half the length of the calyx; calyx-teeth oblong-lanceolate, equalling the tube of the corolla; corolla small, yellowish white, its lobes ovate or triangular-lanceolate, pubescent externally, especially at the tip, with white silky hairs; stamens and styles exserted; upper part of ovary pubescent with white silky hairs.

A handsome little plant, remarkable for its absence of proper leaves. Ashy in color, due to the scale-like covering and pubescence on the stems. Found on the "Flats" at Corpus Christi, on the east side of the San Antonio and Aransas Pass Railroad, a short distance beyond the freight station. In March, when the plants were only an inch or two high, they were found infested with a fungus, *Æcidium Cressae*.

May 31 (1811).

CUSCUTACEAE.

CUSCUTA L. Sp. Pl. 124 (1753).

Cuscuta arvensis Beyrich; Hook. Fl. Bor. Am. 2: 77 (1834), as synonym.

In sand on the beach at the Oso, on *Lycium Carolinianum*, *Lepidium Virginicum*, and other low plants, growing in a thick, tangled mass.

April 12 (1549), type locality, "N. W. America."

POLEMONIACEAE.

PHLOX L. Sp. Pl. 151 (1753).

Phlox Drummondii Hook. Bot. Mag. *t. 3441* (18   ).

In rich soil about Kerrville, especially in damp places. A plant collected in sand along Nueces Bay, called *P. Drummondii villosissima* is apparently only a form of the species. When growing directly in the sand it was weak procumbent, with stems almost two feet long and viscid pubescent, but plants growing only a few feet distant in sod were erect, only three or four inches high and much less pubescent (1435).

April 24 (1641).

GILIA R. & P. Prodr. Fl. Per. 25, *t. 4* (1794).

Gilia rubra (L.).
Polemonium rubrum L. Sp. Pl. 163 (1753).
Cantua coronopifolia Willd. Sp. Pl. 1: 879 (1798).
Gilia coronopifolia Pers.; Lindl. Bot. Reg. *t. 1691* (   ).

On low, stony ground on the left bank of the Guadalupe above Kerrville, altitude 1600 feet; scattered in growth. About half way between Kerrville and San Antonio it was seen growing in large patches.
June 16 (1869).

Gilia rigidula Benth.; D.C. Prodr. 9: 312 (1845).

In dry, usually stony ground, about Kerrville, altitude 1600–1800 feet. As noted by Lindheimer, the flower opens only in the afternoon while the sun is shining.
April 24 (1646); type locality, Texas.

HYDROPHYLLACEAE.

PHACELIA Juss. Gen. Pl. 127 (1789).

Phacelia congesta Hook. Bot. Mag. t. 3452 ( ).
Phacelia conferta Don. Gen. Syst. 4: 397 (1838).

In low, dry ground along Corpus Christi and Nueces Bays, sea level to 20 feet.
March 12 (1432).

Phacelia patuliflora (Engelm. & Gray) A. Gray, Proc. Am. Acad. 10: 321 (1875).
Eutoca patuliflora Engelm. & Gray, Bost. Jour. Nat. Hist. 5: 253 (1845).

In rich soil in shade at Corpus Christi, altitude 10–40 feet.
March 12 (1446); type locality, " woods near San Felipe."

MARILAUNIDIUM Kuntze, Rev. Gen. Pl. 434 (1891).

Marilaunidium hispidum (A. Gray) Kuntze, Rev. Gen. Pl. 434 (1891).
Nama hispidum A. Gray, Proc. Am. Acad. 5: 339 (1862).
Nama Jamaicansis Engelm. & Gray, Bost. Jour. Nat. His. 5: 226 (1845), not L.

In dry, open ground along Wolf Creek, Kerr county, altitude 1800 feet. Very little of it was observed.
May 8 (1725); type locality, "near the Brazos."

Marilaunidium undulatum (H.B.K.) Kunte, Rev. Gen. Pl. 434 (1891).
Nama undulatum H.B.K. Nov. Gen. 3: 130 (1818).

Plentiful at Corpus Christi, in low dry ground, especially within the enclosure of the "Bluff City Park."
March 20 (1461); type locality, near the City of Mexico.

BORAGINACEAE.

EHRETIA P. Br. Civ. & Nat. Hist. Jam. 168 (1755).

Ehretia elliptica DC. Prodr. 9: 503 (1845).

About Corpus Christi, altitude 15-40 feet. Usually a bush or small tree, but occasionally a tree 30 feet high and over a foot in diameter. Often planted at Corpus Christi.

March 26 (1502); type locality, "Mexico," collected by Berlandier.

HELIOTROPIUM L. Sp. Pl. 130 (1753).

Heliotropium Curassavicum L. Sp. Pl. 130 (1753).

Very plentiful in the "Flats" at Corpus Christi, at sea level, growing ing in dense tufts, with stems 1-3 feet long.

April 2 (1516); type locality, "in Americae calidioris maritimis."

Heliotropium tenellum (Nutt.) Torr.; Marcy's Rep. Expl. Red river, 304, *t. 14* (1853).

Lithospermum tenellum Nutt. Trans. Am. Phil. Soc. (II.) **5**: 189 (1837.)

On summits of ridges about Kerrville, altitude 2000 feet; abundant.

June 18 (1875); type locality, prairies of the Red river.

LAPPULA Moench, Meth. 416 (1794).

[ECHINOSPERMUM Sw.; Lehm. Asperit. 113 (1818).]

Lappula Texana (Scheele) Britton, Mem. Torr. Club, **5**: 273 (1894).

Cynoglossum pilosum Nutt. Gen. **1**: 114 (1818), not R. & P.

Echinospermum Texanum Scheele, Linnaea, **25**: 260 (1852).

Echinospermum Redowskii var. *cupulatum* A. Gray, in Brew. -- Wats. Bot Cal. **1**: 530 (1876).

Lappula pilosa A. S. Hitchc. Spring. Fl. Manhattan 30 (1894).

At San Antonio along the tracks of the S. P. R. R., altitude 600 feet. April 17 (1585); from the type locality.

ONOSMODIUM Michx. Fl. Bor. Am. 1: 132 (1803).

Onosmodium Bejariense DC. Prodr. **10**: 70 (1846).

In rich, shaded ground on a hillside along Bear Creek, and in a similar situation four miles north of Kerrville, altitude 1800 feet, were collected a few plants referred to this species. It may be a distinct species, but until more material comes to hand it cannot well be separated.

April 30 (1682); type locality, North Mexico, near Bejar.

Onosmodium Carolinianum (Lam.) A. DC. Prodr. 10: 70 (1846).
Lithospermum Carolinianum Lam. Tabl. Encycl. 1: 367 (1791).
At San Antonio on the grassy left bank of the river, altitude 600 feet. Not plentiful.
May 5 (1702).

VERBENACEAE.

VERBENA L. Sp. Pl. 18 (1753).

Verbena bipinnatifida Nutt. Jour. Acad. Phila. 2: 123 (1821).

Abundant at Corpus Christi and vicinity, principally in dry open ground forming large patches.

March 5 (1385); type locality, "hills of Red river."

Verbena canescens H.B.K. Nov. Gen. 2: 274, *t. 136* (1817).

On the summits of the stony limestone hills about Kerrville, altitude 2000 feet. Not much of it collected.

May 14 (1732); type locality, mountains of Mexico, near Guanaxuato.

Verbena officinalis L. Sp. Pl. 18 (1753).

At Corpus Christi, growing in dry open ground, sea level to 20 feet, was a plant referred to this species. The flowers are usually twice as large, as in ordinary *V. officinalis*, the plant stouter and more simple, with a somewhat different leaf.

March 9 (1419); type locality, "in Europae mediterraneae ruderatis."

Verbena quadrangulata n. sp.

(PLATE 6.)

Herbaceous, prostrate and spreading from an apparently perennial, slender root; whole plant pilose, especially the stems; the upper surface of the leaves smoother, with the pubescence appressed; cymosely branching, leafy throughout; leaves an inch or less in length, opposite, rather distant, broadly ovate, abruptly contracted into a margined petiole, three parted, the middle lobe largest, three to five cleft, the lateral ones two to three cleft; spikes dense, bracts narrowly lanceolate, slender, pointed, slightly more than half the length of the calyx; calyx prominently five-ridged, the ridges green, ending in short, slender tips, which are joined by a scarious connective; flowers very small, white, or pinkish tinged, tube exserted, the spreading lobes notched; fruit shorter than the calyx, four-lobed, the lobes oblong, blunt, from a broad base, angled and pitted, surmounted by a four-winged crown, the wings of which are alternate with the lobes.

This most peculiar plant has the habit and general appearance of the species of *Verbenas* which bear rather large and showy flowers, as *V. bipinnatifida* and *V. Aubletia*, but its very small flower at once throws it out of that group. Its most striking feature is the shape of the fruit, which is well shown in the accompanying plate. The crown of this curiously-shaped seed is much like the four-angled fruit of certain species of *Gaura* and is hollow. Until the exact limit of *Verbena* seeds be known, it is provisionally placed under this genus, although it is probably an undescribed genus.

Collected at Corpus Christi, where it is plentiful in open ground, at both the southern and northern ends of the town. Altitude 10-35 feet.

March 5 (1388).

LIPPIA L. Sp. Pl. 633 (1753).

Lippia nodiflora (L.) Michx. Fl. Bor. Am. 2: 15 (1803).

Verbena nodiflora L. Sp. Pl. 20 (1753).

Abundant about Corpus Christi, in rich, dry ground, altitude sea level to 40 feet. Perennial from a woody base, and large, thick root. Branches creeping extensively and rooting at the nodes. A broader-leaved form (1920), was collected at Kerrville, growing·on wet limestone rocks. Distributed as *L. lanceolata.*

May 30 (1906); type locality, "in Virginia."

LANTANA L. Sp. Pl. 626 (1753).

Lantana Camara L. Sp. Pl. 627 (1753).

This handsome, shrubby species is rather common at Corpus Christi and in the surrounding country, altitude sea level to 40 feet.

March 5 (1386).

CALLICARPA L. Sp. Pl. 111 (1753).

Callicarpa Americana L. Sp. Pl. 111 (1753).

At San Antonio at the Southern Pacific bridge, growing in rich ground, but not plentiful.

June 9 (1832); type locality, "in Virginia, Carolina."

LABIATAE.

HEDEOMA Pers. Syn. 2: 131 (1807).

Hedeoma acinoides Scheele, Linnaea, 22: 592 (1849).

In rich, low ground along streams about Kerrville, usually growing near trees or bushes, altitude 1600–1625 feet.

April 19 (1604); type locality, New Braunfels.

Hedeoma Reverchoni A. Gray.
    Hedeoma Drummondii var. *Reverchoni* A. Gray, Syn. Fl. 2, Part 1, 363 (1878).
    On the steep, stony left bank of the Guadalupe at Kerrville, where it was plentiful; altitude 1635 feet.
    April 27 (1883); type locality, Texas.

SALVIA L. Sp. Pl. 23 (1753).

Salvia azurea Lam. Jour. d'Hist. Nat. 1: 409 ( ).
Salvia acuminatissima Venton. Hort. Cels. 50, *t. 50* (1800).
Salvia angustifolia Mich. Flor. Bor. Am. 1: 15 (1803).
    This species was found sparingly along the stony banks of the Guadalupe and Town Creek about Kerrville, altitude 1600 feet. Perhaps plentiful, but just coming into bloom.
    June 22 (1905).

Salvia ballotaeflora Benth. Lab. Gen. & Sp. 270 (1833).
    Plentiful among the chapparral around Corpus Christi. A brittle, stiff bush, with whitish bark. The pale blue, rather large flowers have a tendency to turn brown in drying, but perhaps this would not happen if the weather were favorable.
    March 5 (1381); type locality, near Toliman, in Mexico.

Salvia farinacea Benth. Lab. Gen. & Sp. 274 (1833).
    This plant is very abundant at Kerrville, about the streets of the town and at medium elevations, 1650–1750 feet. Noticed as far east as Kenedy, Carnes county.
    April 19 (1617); type locality, "in Mexico."

Salvia lanceolata Willd. Enum. 37 (1809).
    Salvia trichostemoides Pursh, Fl. Am. Sept. 1: 19 (1814).
    Occurring as a weed in the gutters and along the streets of Kerrville, but observed on the hills; altitude 1650–1800 feet.
    April 25 (1652).

Salvia pentstemonoides Kunth. Ind. Sem. Berol. 13 (1848).
    On the left bank of the Guadalupe at Kerrville, on a moist limestone ledge, where there was a group of perhaps fifty plants; altitude, 1625 feet. Apparently a rare species. The deep, dull rose-purple flowers at a short distance look like those of some species of *Pentstemon*.
    June 20 (1894); type locality, west Texas?

Salvia Texana (Scheele) Torr. Mex. Bound. Surv. 2: 132 (1859).
    Salviastrum Texanum Scheele, Linnaea, 22: 584 (1849).
    This is one of the abundant and characteristic plants of the limestone

at medium elevations around Kerrville. It usually grows in thick clumps from a stout root, the pale, blue-purple flowers making masses of color which are noticeable at quite a distance.

April 19 (1635); type locality, near Austin.

MONARDA L. Sp. Pl. 22 (1753).

Monarda citriodora Cerv.; Lag. Nov. Gen. & Sp. 2 (1816).

Monarda aristata Nutt. Trans. Am. Phil. Soc. (II.) 5: 186 (1833–37).

Another plant which is abundant about Kerrville, forming large patches, at an altitude of 1650–1800 feet. Flowers dark rose color. It is abundant as far east as San Antonio, and noticed at intervals between there and Kenedy.

May 18 (1761).

Monarda pectinata Nutt. Jour. Acad. Phila. (II.) 1: 182 (1849).

What appears to be this long-lost and rare species is very plentiful about Corpus Christi, especially southeast of the town. My specimens were collected in the Arroyo. Great quantities of it were noticed along the railroad between Corpus Christi and Kenedy. Neither the herbarium of Columbia College nor the U. S. National herbarium at Washington, contain specimens of *M. pectinata*, but at the Academy of Natural Sciences, Philadelphia, is a small specimen from New Mexico, collected by Fendler, No. 602, which is referable to it.

My specimens are rather tall, usually 2 feet and over in height, especially if growing in rich, shaded ground, with a hard, woody rootstock, which gives it the appearance of being a perennial. The flowers are lemon-yellow, resinous dotted, ciliate pubescent. The ciliate bracts are whitened, yellowish, or faintly reddish tinged. The following is Nuttall's original description :

"Biennial? slightly pubescent, leaves oblong, lanceolate, denticulate, shortly petiolate; capituli proliferous, rather small, subtended by herbaceous bracts, some of them purplish, ovate, acute, strongly ciliate, as well as the elongated setaceous teeth of the calyx; corolla widely ringent, the tube scarcely exserted beyond the calyx."

May (31 (1810); type locality, Santa Fé, New Mexico, collected by Gambel.

SCUTELLARIA L. Sp. Pl. 598 (1753).

Scutellaria Drummondii Benth. Lab. Gen. & Sp. 441 (1834).

Growing in sand along the beach, southeast of Corpus Christi, quite near the water. A small form (1503), collected March 27. At Kerrville, where it was found sparingly at an altitude of 1620-1650 feet it was larger and more vigorous.

April 19 (1613); type locality, on the Brazos.

Scutellaria resinosa Torr. Ann. Lyc. N. Y. 2: 232 (1827).
Scutellaria Wrightii A. Gray, Proc. Am. Acad. 8: 370 (1872).

This plant was plentiful about Kerrville, in dry, stony ground, altitude 1620-1800 feet. The deep, blue-purple flowers are very handsome. This plant, as pointed out by Dr. Porter in the Bulletin of the Torrey Club, 21: 176 (1894), has been going under the name of *S. Wrightii*.

April 19 (1606); type locality, on the Canadian.

PHYSOSTEGIA Benth. Lab. Gen. & Sp. 504 (1834).

Physostegia Virginiana (L.) Benth. Lab. Gen. & Sp. 504 (1834).
Dracocephalum Virginianum L. Sp. Pl. 594 (1753).
Dracocephalum speciosum Sweet, Brit. Fl. Gard. *t. 93* (1825).

This plant, called *P. Virginiana* var *speciosa* in the Synoptical Flora, was found sparingly on the banks of the Guadulupe in moist ground. I have now collected it in Virginia, North Carolina and Texas, and have always considered it distinct. Its manner of growth is different from the northern *P. Virginiana*, and the flowers are larger, more inflated and more terminal.

June 22 (1906); type locality, North America.

BRAZORIA Engelm. & Gray, Bost. Jour. Nat. Hist. 5: 255 (1845).

Brazoria scutellarioides Engelm. & Gray, Bost. Jour. Nat. Hist. 5: 257 (1845).
Physostegia truncata Hook. Bot. Mag. *t. 3494* (   ).

Abundant in rich ground on the summits of hills about Kerrville, altitude 2000 feet. A handsome plant, resembling a *Physostegia*.

May 14 (1733); type locality, "near Cat Spring, west of the Brazos."

STACHYS L. Sp. Pl. 580 (1753).

Stachys Drummondii Benth. Lab. Gen. & Sp. 551 (1834).

Growing in rich ground, under trees and bushes, near the upper end of Nueces Bay. Not observed at any other place.

March 12 (1434); type locality, on the Brazos.

TEUCRIUM L. Sp. Pl. 562 (1753).

Teucrium Canadense L. Sp. Pl. 564 (1753).
Teucrium Virginicum L. Sp. Pl. 564 (1753).

In rich, moist ground along Town Creek at Kerrville, altitude 1600 feet. On some plants the bracts are unusually long.
June 16 (1873); type locality, "in Canada."
Teucrium Cubense L. Mant. 80 (1767).
Along the shores of Corpus Christi and Nueces Bays, usually growing in clumps under bushes.
March 12 (1439); type locality, in Cuba.
Teucrium laciniatum Torr. Ann. Lyc. N. Y. 2: 231 (1828).
Growing in patches along the streets at Kerrville, and also on the road between Kerrville and Fredericksburg, altitude 1650-1800 feet.
May 8 (1718); type locality, "on the Rocky Mountains."

SOLANACEAE.
LYCIUM L. Sp. Pl. 191 (1753).
Lycium Carolinianum Walt. Fl. Car. 84 (1788).
On the salt flats at Corpus Christi, where it is abundant, but very little found in flower.
March 6 (1395); type locality, Carolina.

CHAMAESARACHA A. Gray; Benth. & Hook. Gen. Pl. 2: 891 (1876).
Chamaesaracha Coronopus (Dunal) A. Gray, Syn. Fl. 2: Part 2, 232 (1878).
Solanum Coronopus Dunal, in DC. Prodr. 13: Part 1, 64 (1852).
Rather common in dry, open ground about Kerrville, altitude 1650-1700 feet.
April 24 (1647); type locality, Mexico between Laredo and Bejar.

PHYSALIS L. Sp. Pl. 182 (1753).
Physalis mollis Nutt. Trans. Am. Phil. Soc. (II.) 5: 194 (1837).
This plant is rather common at Corpus Christi, usually growing in rich, shaded ground, from sea level to 40 feet.
March 17 (1453); type locality, sandy banks of the Arkansas.
Physalis*
On the bluff southeast of Corpus Christi, in open ground once broken for a street, was a patch of plants with large, pale yellow flowers, rarely with a darker center. The plants were prostrate from a large and thick, fleshy root. The leaves are dull green, rather thick.
March 27 (1507).
Physalis*

---

* These two species are in the hands of Mr. P. A. Rydberg, for determination.

A few specimens of a tall, but rather weak plant, were collected along Town Creek, at Kerrville, in rich, shaded ground. The flower is small, greenish-yellow, the inflated calyx large, and rather slender pointed.
May 17 (1756).

SOLANUM L. Sp. Pl. 184 (1753).

Solanum elaeagnifolium Cav. Ic. 3: 22, *t. 243*.(1794).

A very common plant of the coast region, and apparently everywhere throughout southern Texas. Collected at Corpus Christi, sea level to 40 feet.

March 27 (1511).

Solanum rostratum Dunal, Sol. 234, *t. 24* (1813).

Solanum heterandrum Pursh, Fl. Am. Sept. 156, *t. 7* (1814).

Common in the streets of Kerrville, and also on the hills in rich ground, altitude 1650–1800 feet.

May 17 (1755); described from cultivated specimens by Dunal.

Solanum Torreyi A. Gray, Proc. Am. Acad. 6: 44 (1862).

At San Antonio this species was plentiful in rich ground along the river bank.

May 3 (1709); type locality, "upper Arkansas to lower Texas."

Solanum triquetrum Cav. Ic. 3: 30, *t. 259* (1794).

At Corpus Christi from sea level to 40 feet, usually woody at base; sometimes long and vine-like.

March 6 (1399); type locality, Mexico.

CESTRUM L. Sp. Pl. 191 (1753).

Cestrum Parqui L'Her.

A single clump, or rather hedge, of this species was found at San Antonio, along the roadside. The bushes were tall and slender, growing very closely together.

May 5 (1797).

NICOTIANA L. Sp. Pl. 180 (1753).

Nicotiana repanda Willd.; Lehm. Hist. Gen. Nicot. 40, *t. 3* (1818).

A common plant at Corpus Christi from sea level to 40 feet. The flowers expand in the evening just before dark, closing early the next morning.

March 6 (1498); type locality, "in Cuba."

PETUNIA Juss. Ann. Mus. Paris, 2: 215, *t. 47* (1803).
Petunia parviflora Juss. Ann. Mus. Paris, 2: 216, *t. 47* (1803).
Nicotiana parviflora Lehm. Hist. Gen. Nicot. 48 (1818).
Found growing in depressions at Corpus Christi, from sea level to 40 feet. Very abundant near the upper part of the Arroyo in a water hole. March 9 (1412).

SCROPHULARIACEAE.

VERBASCUM L. Sp. Pl. 177 (1753).
Verbascum Thapsus L. Sp. Pl. 177 (1753).
A few plants were found on the right bank of the Guadalupe, about a mile below Kerrville, altitude 1600 feet. The second station recorded for the species in Texas.
June 22 (1907); type locality, Europe.

LINARIA Juss. Gen. Pl. 120 (1789).
Linaria Canadensis (L.) Dumont, Bot. Cult. 2: 96 (1802).
Antirrhinum Canadense L. Sp. Pl. 618 (1753).
Occurring occasionally in cultivated land near Corpus Christi.
March 12 (1445); type locality, "in Virginia, Canada."

ANTIRRHINUM L. Sp. Pl. 612 (1753).
Antirrhinum antirrhiniflora (Willd.).
Maurandia antirrhiniflora Willd. Enum. Berol. *t. 83* (1816).
Usteria antirrhiniflora Poir. in Lam. Encycl. Suppl. 5: 405 (1817).
Antirrhinum maurandioides A. Gray, Proc. Am. Acad. 7: 376 (1868).
Ipomoea Neallyi Coulter, Cont. U. S. Nat. Herb. No. 2, 46 (1890).
First noticed in cultivation at Corpus Christi, and afterwards found twining over bushes along the upper end of the bay. The handsome purple flowers do not in the least resemble those of an *Ipomoea*, but plainly belong to a Scrophulariaceous plant. The capsule, though, bears a superficial resemblance to that of an *Ipomoea*.
May 29 (1790): type locality, Mexico.

PENTSTEMON Soland. in Ait. Hort. Kew. 3: 511 (1789).
Pentstemon Cobaea Nutt. Trans. Am. Phil. Soc. (II.) 5: 182 (1834).?

The plant doubtfully referred to this species is plentiful about Kerrville, at an altitude of 1600–1800 feet. It varies in height from one to two feet, has large and rather broad flowers, ranging in color from white tinged with blue to almost rose color. Its leaves are inclined to be narrower, and the flower usually not more than half the size of the *P. Cobaea* found further north. There are a number of specimens from Texas in the Herbarium of Columbia College identical with my plants.

April 19 (1610); type locality, Arkansas, on the Red river.

Pentstemon Guadalupensis n. sp.
(PLATE 7.)

Low, 8–15 inches high, branching from the perennial rootstock, which sends down numerous, thick, fibrous roots; glabrous below, the inflorescence glandular pubescent and viscid; root leaves linear or spatulate linear, sessile, clustered, 2–4 inches long, acute or acutish; stem leaves from linear to lanceolate, sessile, becoming broader and shorter as they ascend, the upper with broad, almost cordate base, 1–3 inches long, acute, smooth on both sides, entire, or some of the upper ones sparingly denticulate, prominently one-nerved; calyx-teeth about the length of the corolla tube, lanceolate or ovate-lanceolate, glandular puberulent, especially on the margins; corolla white or sometimes faintly tinged with purple, short, less than an inch in length, broad in proportion, the spreading lobes almost equal; sterile filament broadened above, the upper half bearded on one side with yellow hairs.

This species belongs in the Genuini division, near *P. tubiflorus* and *P. albidus*. It was distributed under the latter name, as there are specimens in the National Herbarium identical with my plants, which are called *P. albidus*. It is very plentiful in dry, stony ground along the Guadalupe and Town Creek, altitude 1600–1650 feet, and often growing in company with *P. Cobaea?* Sometimes large patches of ground are white with it. Usually several plants grow together in a clump.

April 19 (1609).

Pentstemon triflorus n. sp.
(PLATE 8.)

Herbaceous, erect, usually 2–3 feet high, simple, very glabrous up to the inflorescence; root leaves spatulate, on margined petioles about equal in length to the blade, entire or minutely denticulate, obtuse or acutish; stem leaves entire or dentate, the dentations sometimes present only on the lower part, sometimes near the upper end, and occasionally most marked in the middle; the first two or three pairs oblong, on rather short, margined petioles, the others sessile, becoming shorter and broader as they ascend the stem, the upper pair especially, which are usually ovate lanceolate and clasping; inflorescence glandular puberulent and viscid; peduncles slender, less than an inch in length, usually three flowered; pedicels shorter than the calyx as a rule; calyx equalling or slightly exceeding the slender tube, the slender, oblong or lanceolate lobes glandular ciliate; flowers bright rose-purple, paler inside and marked with dark stripes, an inch or more in length, gradually dilated, the lobes spreading; sterile filament smooth, slightly dilated at the tip.

Some leaf forms of this striking plant are much like those of the plant referred to *P. Cobaea*, but are thinner, dark green and shining; besides, the flowers are more slender and longer. The color of the flowers is much like that of the beautiful *P. Smallii* of the Carolina mountains. It was found only along the summit of one hill northeast of Kerrville, altitude 2000 feet, an elevation not reaches by the two other species.
April 26 (1654).

CONOBEA Aubl. Pl. Guian. **2**: 639, *t. 258* (1775).

Conobea multifida (Michx.) Benth. in DC. Prodr. **10**: 391 (1846).
Capraria multifida Michx. Fl. Bor. Am. **2**: 22, *t. 35* (1803).

In moist, sandy ground, right bank of the Guadalupe, at Kerrville, altitude 1600 feet; plentiful.

July 2 (1926); type locality, Tennessee and Illinois.

MONNIERA P. Br. Civ. & Nat. Hist. Jam. 269, *t. 28, f. 3* (1755).

Monniera Monniera (L.) Britton, Mem. Torr. Club, **5**: 292 (1894).
Gratiola Monniera L. Cent. Pl. **2**: (1756).
Limosella calycina Forsk. Fl. Æg. Arab. 112 (1775).
Herpestis cuneifolia Pursh, Fl. Am. Sept. 418 (1814).
Herpestis Monniera H.B.K. Nov. Gen. **2**: 366 (1817).
Monniera calycina Kuntze, Rev. Gen. Pl. 462 (1891).

Along the beach at Corpus Christi in wet sand, growing in thick mats. June 5 (1823).

Monniera procumbens (Mill.) Kuntze, Rev. Gen. Pl. 463 (1891).
Erinus procumbens Mill. Dict. (1768).
Herpestis chamaedryoides Benth.

Scattered about Corpus Christi in rich ground, sea level to 40 feet;

near Gregory, San Patricio county, 35 feet, and at Kenedy, Carnes county, 400 feet.
March-June (1460).

VERONICA L. Sp. Pl. 9 (1753).
Veronica peregrina L. Sp. Pl. 14 (1753).
A few plants found in a moist gutter at Kerrville, altitude 1650 feet.
May 17 (1758); type locality, Europe.

CASTILLEJA Mutis; L. f. Suppl. 47 (1781).
Castilleja Lindheimeri A. Gray, Syn. Fl. 2: Part 1, 298 (1878).
Castilleja purpurea A. Gray, Am. Jour. Sci. (II.) 34: 338 (1862), not Don & Benth.
A common plant on the hillsides about Kerrville, altitude 1650-1900 feet. The flowers are somewhat variable in color, but usually scarlet.
April 23 (1630); from type locality, but probably further east or south.

LENTIBULARIACEAE.

UTRICULARIA L. Sp. Pl. 18 (1753).
Utricularia biflora Lam. Tabl. Encycl. 1: 50 (1791).
Utricularia longirostris LeConte, Ann. Lyc. N. Y. 1: 76 (1824).
In mud and water on the right bank of the Guadalupe at Kerrville, altitude 1600 feet.
July 2 (1941).

ACANTHACEAE.

CALOPHANES Don in Sweet, Brit. Fl. Gard. (II.) *t. 181* (1833).
Calophanes linearis (T. & G.) A. Gray, Syn. Fl. 2: Part 1, 324 (1878).
Dipteracanthus linearis T. & G. Bost. Jour. Nat. Hist. 5: 258 (1845).
Plentiful in grassy land along Town Creek at Kerrville. First noticed at the Oso, where only two plants were found. The corolla is a pale purplish-blue.
April 9 (1529); type locality, Texas.

RUELLIA L. Sp. Pl. 634 (1753).
Ruellia clandestina L. Sp. Pl. 634 (1753).
Ruellia tuberosa L. Sp. Pl. 636 (1753).

At Corpus Christi, ranging from sea level to 40 feet. Only one or two plants were found in flower, all the others being in fruit. The flowers were blue.

March 17 (1417); type locality, Barbados.

SIPHONOGLOSSA Œrsted.

Siphonoglossa dipteracantha (Nees).
Adhatoda dipteracantha Nees, in DC. Prodr. 11 : 396 (1847).
Monechma Pilosella Nees, in DC. Prodr. 11 : 412 (1847).
Siphonoglossa Pilosella Torr. Mex. Bound. Surv. 2 : 134 (1859).

Usually under chapparral, or in other protected situations at Corpus Christi, altitude sea level to 40 feet.

March 12 (1425); type locality, Mexico.

DIANTHERA L. Sp. Pl. 27 (1753).

Dianthera Americana L. Sp. Pl. 27 (1753).

Found growing in Town Creek at Kerrville, altitude 1600 feet. Rather plentiful.

May 15 (1748); type locality, Virginia.

PLANTAGINACEAE.

PLANTAGO L. Sp. Pl. 112 (1753).

Plantago aristata Michx. Fl. Bor. Am. 1 : 95 (1803).
Plantago Patagonica var. *aristata* A. Gray, Man. Ed. 2, 269 (1856).

A slender form referred to this species was found growing in a yard at Kerrville, altitude 1650 feet.

May 19 (1769); type locality, Illinois.

Plantago heterophylla Nutt. Trans. Am. Phil. Soc. (II.) 5 : 177 (1833-37).

Rather common on the plateau at Corpus Christi, the plants large and well developed in a moist place, small and stunted in dry ground.

March 17 (1456); type locality " on the banks of the Mississippi and Arkansas.

Plantago Mexicana Link, Enum. Hort. Berol. 1 : 121 (1821).

A spreading, prostrate form of this species (1392), with rather narrow leaves was found on the sloping banks of the bluff at the southeastern end of Corpus Christi, and later an abundance of it was seen on hilltops about Kerrville, altitude 1900-2000 feet. This species, usually confused with *P. Purshii* (*gnaphalioides*), is distinguished from that species by its short and smoother heads.   The plants from Kerrville were always erect.

April 24 (1649); type locality, Mexico.

Plantago Virginica L. Sp. Pl. 113 (1753).

A rather peculiar form of this species was collected in the shell deposit at Corpus Christi, and along the railroad embankments. The plants are small, prostrate, with a very short scape; leaves rather thick, with distant, sharp teeth.

March 8 (1410); type locality, Virginia.

RUBIACEAE.

HOUSTONIA L: Sp. Pl. 105 (1753).

Houstonia angustifolia Michx. Fl. Bor. Am. 1: 85 (1803).

Oldenlandia angustifolia A. Gray, Pl. Wright. 2: 68 (1853).

Abundant at Kerrville from the lowest elevations to about 1800 feet, but not found on the upper slopes and summits of the hills.

April 27 (1661); type locality, "in submaritimis Florida."

Houstonia salina n. sp.

(PLATE 9.)

Prostrate, from an apparently perennial root; stems usually about four inches in length, sometimes six or eight inches, cymosely branched, more or less winged, glabrous; leaves sessile, thick, oblong, slightly narrowed at each end, acutish, the margins rolled in on the under side, glabrous; stipules more or less fimbriate; peduncles usually longer than the leaves just beneath them; pedicels short, two or three flowered, calyx-teeth triangular-ovate or ovate-lanceolate, short; flowers funnel-form, about the size of those of *H. angustifolia*, usually dense, white or pinkish; inner face of the corolla lobes puberulent or glandular, especially on the margins, the throat pubescent; pod one-fourth free.

This species, in some respects resembling *H. angustifolia*, is yet very different from it in both habit and habitat, and in several other respects. Found only in the shell deposit along the beach at Corpus Christi, growing in prostrate tufts.

May 31 (1812).

CEPHALANTHUS L. Sp. Pl. 95 (1753).

Cephalanthus occidentalis L. Sp. Pl. 95 (1753).

In moist places along Town Creek and the Guadalupe at Kerrville, altitude 1600 feet. Rather common. Sometimes almost 10 feet high.
June 16 (1871); type locality, North America.

CRUSEA Cham. Linnaea, 5 : 165 (1830).
Crusea tricocca (T. & G.).
Diodia tricocca T. & G. Fl. N. A. 2: 30 (1841).
Diodia tetracocca Hemsley, Diag. Pl. Nov. 1: 32, *t. 40, f. 10–15*
( ).
Crusea allococca A. Gray, Proc. Am. Acad. 19 : 78 (1884).

Found at Corpus Christi, especially in moist ground, altitude 40 feet, and near Gregory, San Patricio county, altitude 35 feet, in open, bare places in pasture land. When imposing the specific name of *allococca*, Dr. Gray remarks: " Referring this to *Crusea*, I shall not add unnecessarily to synonomy by imposing this spe ific name." The giving of another name to a plant already characterized is always an unnecessary addition to synonomy, and much more so when the author knows that he is adding another name, simply because the older one is a poor one.
April 14 (1573); type locality, Texas.

GALIUM L. Sp. Pl. 105 (1753).
Galium spurium L. Sp. Pl. 106 (1753).
Galium Vaillantii DC. Fl. France, 4: 263 (1805).
Galium Aparine var. *Vaillantii* Koch, Fl. Germ. 330 (1837).

At San Antonio, in shaded ground, along the Southern Pacific Railroad.
April 17 (1588); type locality, Europe.

Galium Texense A. Gray, Proc. Am. Acad. 19: 80 (1884).
Common in rich ground about Corpus Christi. Old plants, large and spreading, prostrate; altitude, sea level to 40 feet.
March 12 (1437); type locality, Texas.

CAPRIFOLIACEAE.

VIBURNUM L. Sp. Pl. 267 (1753).
Viburnum prunifolium L. Sp. Pl. 268 (1753).
Occasional in rich ground in the vicinity of Kerrville, altitude 1620–1800 feet. A bush 6–10 feet high.
April 19 (1595); type locality, Virginia and Carolina.

SYMPHORICARPOS Juss. Gen. 211 (1789).

Symphoricarpos Symphoricarpos (L.) MacM. Bull. Torr. Club, **19**: 15 (1892).
  Lonicera Symphoricarpos L. Sp. Pl. 175 (1753).
  Symphoricarpos orbiculatus Moench. Meth. 503 (1794).
  Symphoricarpos vulgaris Michx. Fl. Bor. Am. **1**: 106 (1803).

In a copse in low ground near Waco, McLennan county. Collected only in fruit, without leaves.

March 2 (1373); type locality, Virginia and Carolina.

LONICERA L. Sp. Pl. 173 (1753).

Lonicera albiflora T. & G. Fl. N. A. **2**: 6 (1841).
  Lonicera dumosa A. Gray, Pl. Wright. **2**: 66 (1853).

As a vine, hanging from the dripping limestone bluff on the left bank of the Guadalupe at Kerrville, and at the base of the hills, northwest of the town, when it was a shrub, growing in thickets.

April 19 (1597); type locality, "prairies near Ft. Towson, on the Arkansas."

VALERIANACEAE.

VALERIANELLA Poll. Hist. Pl. Palat. **1**: 29 (1776).

Valerianella amarella (Lindheimer) Krok. Monog. Valer. in Svenska Vetensk Acad. Handl. **5**: No. 1 (1864).
  Fedia amarella Lindheimer; Engelm. in A. Gray, Bost. Jour. Nat. Hist. **6**: 217 (1850).

A few slender plants were found in rich, low, shaded ground on the right bank of the Guadalupe, at Kerrville. On some of the hilltops it is extremely plentiful, occurring as a low and spreading plant, the small white flowers exhaling a pleasant odor, which is quite marked when one approaches a patch of plants. The odor of the dried plant is very much like that of tincture of valerian.

April 19 (1623); type locality, Comanche Spring, Texas.

CUCURBITACEAE.

CUCURBITA L. Sp. Pl. 1010 (1753).

Cucurbita foetidissima H. B. K. Nov. Gen. **2**: 123 (1817).
  Cucumis perennis James, in Long's Exp. **2**: 20 (1823).
  Cucurbita perennis A. Gray, Bost. Jour. Nat. Hist. **6**: 193 (1850).

Rather common about Kerrville, altitude 1650-1800 feet. Also noticed along the railroad east of San Antonio. Stems often ten or fifteen feet long.

May 8 (1727); type locality, Mexico, near Guanaxuato.

MAXIMOWICZA Cogn.; DC. Monog. Phan. 3: 726 (1881).
Maximowicza Lindheimeri (A. Gray) Gogn. ; DC. Monog. Phan. 3: 727 (1881).
Sicydium Lindheimeri A.Gray, Bost. Journ. Nat. Hist. 6: 194 (1850).

In low, rich ground about Kerrville, climbing over bushes; altitude 1600 feet.

May 3 (1694); type locality, "thickets, New Braunfels to the Liano," Texas.

CAMPANULACEAE.

LEGOUZIA Durand, Fl. Bourg. 2 : 26 (1782).
[SPECULARIA Heist. ; A. DC. Mon. Camp. 344 (1830).]
[PENTAGONIA Sieg. ; Kuntze, Rev. Gen. Pl. 381 (1891).]
Legouzia biflora (R. & P.) Britton, Mem. Torr. Club, 5 : 309 (1894).
Campanula biflora R. & P. Fl. Per. 2: 55, *t. 200, f. 6* (1799).
Specularia biflora A. Gray, Proc. Am. Acad. 11 : 82 (1876).

Collected in grassy ground on the plateau, six miles north of Kerrville, altitude 1900 feet.

May 8 (1721.

Legouzia Coloradoense (Buckley).
Campanula Coloradoense Buckley, Proc. Phila. Acad. 400 (1861).
Specularia Lindheimeri Vatké, Linnaea, 38: 713 (1874).

On the summit of a hill north of Kerrville, where it is plentiful, altitude 2000 feet. The pods on these specimens are usually twisted, which is a character contrary to that give by Coulter in the Manual of Western Texas.

May 14 (1731); type locality, " upper Colorado of Texas."

COMPOSITAE.

VERNONIA Schreb. Gen. Pl. 2: 541 (1791).
Vernonia Drummondii Shuttlw. ; Werner, Jour. Cincin. Soc. Nat. Hist. 16: 171 (1894).
Vernonia altissima var. *grandiflora* A. Gray, Syn. Fl. 1 : Part 2, 90 (1884).

Plentiful in thickets along the banks of the Guadalupe, altitude 1600 feet. Stems five to six feet high. Distributed as *V. Baldwinii*.
June 29 (1927).

Vernonia fasciculata Michx. Fl. Bor. Am. 2 : 94 (1803).
Found sparingly along the Guadalupe below Kerrville, altitude 1600 feet. Low, some forms of it approaching *V. Lindheimeri*. Distributed as *V. Guadalupensis* n. sp.
June 22 (1909) ; type locality, Illinois.

Vernonia Lindheimeri Engelm. & Gray, Proc. Am. Acad. 1 : 46 (1846).
Common in dry stony ground along Town Creek and the Guadalupe, at Kerrville, altitude 1600–1650 feet.
June 22 (1908) ; type locality, New Braunfels, Texas.

EUPATORIUM L. Sp. Pl. 836 (1753).

Eupatorium conyzoides Vahl. Symb. Bot. 3 : 96 (1794).
A few plants were found on the edge of the plateau southeast of Corpus Christi. Apparently not recorded from so far east in Texas, as it is given as "along the Rio Grande."
June 2 (1816).

Eupatorium incarnatum Walt. Fl. Car. 200 (1788).
Two or three stunted plants were found under chapparral bushes at Corpus Christi, altitude 20 feet.
March 6 (1398); type locality, Carolina.

GRINDELIA Willd. Gesell. Nat. Fr. Berlin, Mag. 1 : 260 (1807).

Grindelia inuloides Willd. Gesell. Nat. Fr. Berlin, Mag. 1 : 260 (1807).
Rather common on the plateau southeast of Corpus Christi, altitude 35 feet.
June 2 (1820).

Grindelia squarrosa grandiflora A. Gray, Pl. Wright. 1 : 98 (1852).
Grindelia grandiflora Hook. Bot. Mag. t. *4628* ( ).
Just coming into bloom on hillsides about Kerrville, and rather plentiful at medium elevations, 1650–1800 feet.
June 22 (1904).

CHRYSOPSIS Nutt. Gen. 2 : 150 (1818).

Chrysopsis villosa canescens (DC.) A. Gray, Syn. Fl. 1: Part 2, 123 (1884).
Aplopappus ? (Leucopsis) canescens DC. Prodr. 5 : 349 (1836).

On the edge of the bluff on the left bank of the Guadalupe, at Kerrville, growing in clumps; altitude 1650 feet.
June 13 (1854); .type locality, Texas.

ERIOCARPUM Nutt. Trans. Am. Phil. Soc. (II.) 7: 320 (1841).
Eriocarpum rubiginosum phyllocephalum (DC.).
Aplopappus phyllocephalus DC. Prodr. 5: 347 (1836).
Aplopappus rubiginosus var. *phyllocephalus* A. Gray, Syn. Fl. 1: Part 2, 130 (1884).
Found sparingly in dry ground on the edge of the " Flats," at Corpus Christi, and along Nueces Bay.
March 12 (1440).

Eriocarpum spinulosum (Pursh) Greene, Erythea, 2: 108 (1894).
Amellus spinulosus Pursh, Fl. Am. Sept. 564 (1814).
Aplopappus (?) *spinulosus* DC. Prodr. 5: 347 (1836).
Two clumps of this plant were found in grassy pasture land at Kerrville, altitude 1750 feet.
June 14 (1858).

CHONDROPHORA Raf. New Fl. N. A. 4: 79 (1836).
[BIGELOWIA DC. Prodr. 5: 329 (1836), not Spreng.]
Chondrophora Drummondii (T. & G.).
Linosyris Drummondii T. & G. Fl. N. A. 2: 233 (1842).
Bigelovia Drummondii A. Gray, Proc. Am. Acad. 8: 639 (1873).
On the plateau near the Arroyo at Corpus Christi, and at the Oso, near the water, apparently rather plentiful, but just coming into bloom when I left. A rare plant in herbaria.
April 12 (1557); type locality, coast of Texas.

APHANOSTEPHUS DC. Prodr. 5: 310 (1836).
Aphanostephus humilis (Benth.) A. Gray, Pl. Wright, 1: 93 (1852).
Leucopsidium humile Benth. Pl. Hartw. 18 (1836).
In open, exposed ground at Corpus Christi, in the northern end of the town near the beach. Plants prostrate, small.
March 8 (1404).
Aphanostephus ramosissimus DC. Prodr. 5: 310 (1836).
Aphanostephus Riddellii T. & G. Fl. N. A. 2: 189 (1841).
Along the banks of the Gaudalupe, in dry, stony ground, at Kerrville, altitude 1620 feet. Growing in clumps.
April 24 (1642); type locality, Mexico and Texas.

ERIGERON L. Sp. Pl. 863 (1753).

Erigeron Canadensis L. Sp. Pl. 863 (1753).
Occasional about the streets of Kerrville, altitude 1650 feet.
June 30 (1933); type locality, Canada.

Erigeron Philadelphicus L. Sp. Pl. 863 (1753).
Along Town Creek and the Guadalupe, altitude 1600 feet, in rich, moist, shaded ground, is found a white-flowered form with mostly entire leaves. In the Herbarium of Columbia College there is only one specimen exactly like it, collected also in Texas. It may yet prove to be distinct from our larger, pink-rayed plant of the north.
April 28 (1672).

Erigeron repens A. Gray, Syn. Fl. 1: Part 2, 217 (1884).
Two or three plants picked up at Flower Bluff, growing in the scrub oak in sand.
April 9 (1538); type locality, coast of Texas.

Erigeron tenuis T. & G. Fl. N. A. 2: 175 (1841).
This little plant was found in quantity on a grassy bank near the upper end of Nueces Bay. The ray flowers were tinged with blue.
March 12 (1436), range, from Arkansas to Texas.

FILAGO L. Sp. Pl. 927 (1753).

[EVAX Gaertn. Fr. & Sem. 2: 393, *t. 165 f. 3* (1791).]

Filago multicaulis (DC.).
Evax multicaulis DC. Prodr. 5: 459 (1836).
Filaginopsis multicaulis T. & G. Fl. N. A. 2: 263 (1842).
Very common in the low ground bordering on the "Flats" at Corpus Christi; also on the railroad embankment.
March 14 (1450); type locality, Texas.

GNAPHALIUM L. Sp. Pl. 850 (1753).

Gnaphalium purpureum L. Sp. Pl. 854 (1753).
A few plants were found at Corpus Christi and Flower Bluff, in sand, altitude 15-40 feet.
March 23 (1489); type locality, North America.

MELAMPODIUM L. Sp. Pl. 921 (1753).

Melampodium cinereum DC. Prodr. 5: 518 (1836).
One of the commonest plants about Kerrville, altitude 1620-1800 feet. Growing in bunches.
April 23 (1632); type locality, Mexico.

SILPHIUM L. Sp. Pl. 919 (1753).
Silphium integrifolium Michx. Fl. Bor. Am. 2: 146 (1803).
One patch of it on a dripping limestone ledge, left bank of the Guadalupe, at Kerrville, altitude 1620 feet. Stems 4-6 feet high.
June 20 (1895); type locality, Illinois.
Silphium laciniatum L. Sp. Pl. 919 (1753).
A few plants were found on hillsides at middle elevations about Kerrville, 1700-1850 feet.
June 30 (1930); type locality, North America.

BERLANDIERA DC. Prodr. 5: 517 (1836).
Berlandiera Texana DC. Prodr. 5: 517 (1836).
Occasionally found in low, usually damp ground, along the Guadalupe and Town Creek, at Kerrville.
June 16 (1874); type locality, Texas.

LINDHEIMERA Engelm. & Gray, Proc. Am. Acad. 1: 47 (1846).
Lindheimera Texana Engelm. & Gray, Proc. Am. Acad. 1: 47 (1846).
A common plant about Kerrville, growing along roadsides and in fields, altitude 1700-1900 feet.
April 26 (1660); type locality, New Braunfels, Texas.

ENGELMANNIA T. & G. Fl. N. A. 2: 283 (1841).
Engelmannia pinnatifida T. & G. Fl. N. A. 2: 283 (1841).
First collected along the upper end of Nueces Bay on a grassy bank, later at San Antonio. At Kerrville it is very plentiful in stony ground at an altitude of 1600-1650 feet.
March 12 (1522); range, Arkansas to Texas.

PARTHENIUM L. Sp. Pl. 988 (1753).
Parthenium Hysterophorus L. Sp. Pl. 988 (1753).
Argyrochaeta bipinnatifida Cav. Ic. 4: 54, *t. 378* (1797).
One of the most common weeds at Corpus Christi, San Antonio, and Kerrville. Not observed at any distance from either of the places named.
March-June (1418); type locality, Jamaica.

AMBROSIA L. Sp. Pl. 987 (1753).
Ambrosia artemisiaefolia L. Sp. Pl. 988 (1753).
Several thick bunches of it found along the Southern Pacific Railroad near the bridge, at San Antonio.
May 3 (1696); type locality, Virginia, Pennsylvania.

RUDBECKIA L. Sp. Pl. 906 (1753).

Rudbeckia amplexicaulis Vahl. Act. Havn. 2: 29, *t. 4* (1783).

A few plants were found in wet ground on the edge of the Guadalupe at Kerrville, altitude 1600 feet. The parti-colored flowers of this and the next species are very handsome.

June 13 (1853).

Rudbeckia bicolor Nutt. Jour. Acad. Phila. 7: 81 (1834).

Collected first in San Patricio county (1571), and later at Kerrville along the Guadalupe, altitude 1600 feet, where it was plentiful, in rich, low, shaded ground.

May 19 (1764); type locality, Arkansas, near the Red river.

LEPACHYS Raf. Jour. Phys. 89: 100 (1819).

Lepachys columnaris (Pursh) T. & G. Fl. N. A. 2: 315 (1842).

Rudbeckia columnaris Pursh, Fl. Am. Sept. 575 (1814).

At Kerrville, collected along the Guadalupe and at the base of the hills, altitude 1600-1750 feet; common.

June 13 (1850); type locality, on the Missouri.

Lepachys columnaris pulcherrima T. & G. Fl. N. A. 2: 315 (1842).

Very common in waste places about San Antonio, altitude 600 feet. In the "Check List" Dr. Britton has included this under the species, but I keep it separate here, inasmuch as the living plants of the two are quite different in appearance. At San Antonio, where thousands of plants of this are found growing in clumps, only one yellow-flowered plant was seen. At Corpus Christi, where it is rather common on the plateau, no yellow-flowered ones were seen. It seems to be more gregarious in habit and stouter than true *L. columnaris*.

May 5 (1708).

Lepachys peduncularis T. & G. Fl. N. A. 2: 315 (1842).

Scattered here and there, in low ground along the beach, at the upper end of Corpus Christi Bay. Heads on a scape-like peduncle, leaves all at the base, thick and fleshy. Rays short, brown-purple.

May 29 (1789); type locality, Texas, in low ground.

BRAUNERIA Neck. Elem. 1: 17 (1790).

[ECHINACEA Moench. Meth. 591 (1794).]

Brauneria pallida (Nutt.) Britton, Mem. Torr. Club, 5: 333 (1894).

Rudbeckia pallida Nutt. Jour. Acad. Phila. 7: 77 (1834).

Echinacea angustifolia DC. Prodr. 5: 554 (1836).

Plentiful, but scattered, on hillsides about Kerrville, altitude 1650–1850 feet.
May 14 (1735); type locality, Arkansas.

TETRAGONOTHECA L. Sp. Pl. 903 (1753).
Tetragonotheca Texana Engelm. & Gray, Proc. Am. Acad. 1: 48 (1846).
A few plants of this were found on the low, gravelly right bank of the Guadalupe, at Kerrville, altitude 1600 feet.
June 13 (1848); type locality, junction of the Guadalupe and Cibola.

BORRICHIA Adans. Fam. Pl. 2: 130 (1763).
Borrichia frutescens (L.) DC. Prodr. 5: 489 (1836).
Buphthalmum frutescens L. Sp. Pl. 903 (1753).
Common along the beach of Corpus Christi Bay.
May 29 (1786); type locality, Jamaica and Virginia.

HELIANTHUS L. Sp. Pl. 904 (1753).
Helianthus annuus L. Sp. Pl. 904 (1753).
About the streets of Kerrville and in cultivated fields, altitude 1650–1700 feet.
June 14 (1862); type locality, North America.

ENCELIA Adans. Fam. Pl. 2: 128 (1763).
Encelia calva (Engelm. & Gray) A. Gray, Proc. Am. Acad. 19: 8 (1883).
Barrattia calva Engelm. & Gray, Proc. Am. Acad. 1: 48 (1846).
Scattered, on hillsides about Kerrville, altitude 1700–1900 feet. Flowers pale yellow, rather large.
June 14 (1860); from the type locality, on the upper Guadalupe.

ZEXMENIA Llave & Lex. Nov. Veg. Descr. 1: 13 (1824).
Zexmenia hispida (H.B.K.) A. Gray, Proc. Am. Acad. 19: 10 (1883).
Wedelia hispida H.B.K. Nov. Gen. 4: 215, *t. 371* (1820).
Stemmodontia scaberrima Cassini, Dict. 46: 407 (1826).
Lipochaeta Texana T. & G. Fl. N. A. 2: 357 (1842).
Zexmenia Texana A. Gray, Pl. Wright. 1: 112 (1852).
Viguiera longipes Coulter, Cont. U. S. Nat. Herb. No. 2, 41 (1890).

This plant is common on the plateau near Corpus Christi, especially at the Oso, and is found at Kerrville ranging pretty well up on the hillsides. A handsome species with orange-colored rays. Well figured in Cont. U. S. Nat. Herb. 2: 220 (1891).
March 21 (1479); type locality, Tenochtitlensi, Mexico.

VERBESINA L. Sp. Pl. 901 (1753).
[RIDAN Adans. Fam. Pl. 2: 130 (1763).]
[ACTINOMERIS Nutt. Gen. 2: 181 (1818).]

Verbesina encelioides (Cav.) A. Gray, Syn. Fl. 1: Part. 2, 288 (1884).
Ximenesia encelioides Cav. Icon. 2: 60, *t. 178* (1793).
Common about Corpus Christi, often becoming a weed in cultivated ground.
May 29 (1785); range, Arizona to Texas.

Verbesina Texana Buckley, Proc. Acad. Phila. 458 (1861).
Not infrequent around Corpus Christi. This plant, a very good species, was reduced by Gray to *V. Virginica*, in the rather hasty condemning of the majority of Buckley's Texas species, a great many of which have since been reinstated, and no doubt a number more will be. It is readily separated from *V. Virginica* by its thick, repand leaves, which have very broad petioles, the hemispherical involucre, and broadly winged achene. In the herbarium of Columbia College it is represented by specimens from Miss Mary B. Croft (40) collected at San Diego, Texas, and by Pringle's 1916 from in a valley near Monterey.
April 12 (1556).

SYNEDRELLA Gaertn. Fr. & Sem. 2: 456, *t. 171* (1791).
Synedrella vialis (Less.) A. Gray, Proc. Am. Acad. 17: 217 (1882).
Calyptocarpus vialis Less. Syn. 221 (1832).
Oligogyne Tampicana DC. Prodr. 5: 629 (1836).
A small plant, from a woody base, found about Corpus Christi. The yellow ray flowers are very small.
March 26 (1500); type locality, Mexico.

COREOPSIS L. Sp. Pl. 907 (1753).
Coreopsis Nuecensis n. n.
Coreopsis coronata Hook. Bot. Mag. *t. 3460* (   ), not L. (1753), and not Walt. (1788).
Plentiful about Corpus Christi in moist ground, usually growing in

bunches, the more vigorous plants a foot high or more. It seems that
Hooker was aware that the specific name *coronata* had been previously
used, yet he took it up for this plant, and has been followed ever since.
April 11 (1548).

Coreopsis Drummondii (Don) T. & G. Fl. N. A. 2: 354 (1842).
Calliopsis Drummondii Don, in Sweet's Brit. Fl. Gard. Ser. 2, *t.*
315 ( ).
This elegant species is common on hilltops about Kerrville, where it
grows in company with *Thelesperma trifidum*.
May 14 (1730), type locality, Texas.

THELESPERMA Less. Linnaea, 6: 511 (1831).

Thelesperma simplicifolium A. Gray, Kew Jour. Bot. 1: 252 (1849).
Cosmidium simplicifolium A. Gray, Mem. Am. Acad. 4: 86 (1849).
Thelesperma subsimplicifolinm A. Gray; Torr. Mex. Bound. Surv.
2: 90 (1859).
Rather common in dry, stony ground at Kerrville, altitude 1650-1800
feet.
May 10 (1728); type locality, Buena Vista, Mexico.

Thelesperma trifidum (Poir.) Britton, Trans. N. Y. Acad. Sci. 9:
182 (1890).
Coreopsis trifida Poir. in Suppl. Lam. Encycl. 2: 353 (1811).
Thelesperma filifolium A. Gray, Kew Jour. Bot. 1: 252 (1849).
Plentiful on summits of hills in Kerr county, altitude 1900-2000 feet.
There is considerable resemblance between this plant and *Coreopsis
Drummondii* except in the color of the flowers, yet they can readily be
distinguished when growing side by side.
April 28 (1665); type locality, North America.

MARSHALLIA Schreb. Gen. Pl. 810 (1789).

Marshallia caespitosa Nutt.; DC. Prodr. 5: 680 (1836).
In stony, dry ground along Town Creek and the Guadalupe, altitude
1600-1650 feet, this species is abundant. Flowers pinkish or purplish
tinged.
April 19 (1618); type locality, Red river in Arkansas.

HYMENOPAPPUS L'Hér.; Michx. Fl. Bor. Am. 2: 103 (1803).

Hymenopappus artemisiaefolius DC. Prodr. 5: 658 (1836).
Scattered, but plentiful near Kerrville, from the lowest elevations to
the highest, 1600-2000 feet, but most abundant on the hilltops.
April 23 (1638); type locality, Texas.

FLORESTINA Cass. Bull. Soc. Philom. 175 (1815).

Florestina tripteris DC. Prodr. 5: 655 (1836).

On the edge of the "Flats" at Corpus Christi, at the base of the plateau.

June 8 (1828); type locality, Laredo, Texas.

POLYPTERIS Nutt. Gen. 2: 139 (1818).

Polypteris callosa (Nutt.) A. Gray, Proc. Am. Acad. 19: 30 (1883).

Stevia callosa Nutt. Jour. Acad. Phila. 2: 121 (1821).

An occasional plant was found in the then dry and gravelly beds of Town Creek and the Guadalupe, altitude 1600 feet. Probably plentiful, but just coming into bloom.

June 27 (1919); type locality, "gravelly banks of the Arkansas."

Polypteris Hookeriana (T. & G.) A. Gray, Proc. Am. Acad. 19: 31·(1883)

Palafoxia Hookeriana T. & G. Fl. N. A. 2: 368 (1842).

Growing in sand near the shore at the Oso, and in the shell deposit at the upper end of Corpus Christi.· Rather plentiful at both places.

April 12 (1562); type locality, Texas.

HYMENATHERUM Cass. Bull. Soc. Philom. 1817, 12 (1817).

Hymenatherum tagetoides (T. & G.) A. Gray, Mem. Am. Acad. 4: 88 (1849).

Dysodia tagetoides T. & G. Fl. N. A. 2: 361 (1842).

Found only on the upper slopes of the hills at Kerrville, altitude 1800–1900 feet, growing in open, grassy pasture land. Plants scattered, but plentiful.

June 14 (1855); type locality, Texas.

Hymenatherum Wrightii A. Gray, Mem. Am. Acad. 4: 89 (1849).

At Corpus Christi, near the water tank of the Texas and Mexican Railway, this species grew in dense patches in an inclosed piece of ground. A few plants were also found at the Oso, where they were larger and more vigorous, the situation being more favorable for growth.

March 23 (1494); type locality, Texas.

HELENIUM L. Sp. Pl. 886 (1753).

Helenium elegans DC. Prodr. 5: 667 (1836).

This handsome, though small-flowered species, was found at Kerrville in moist or wet ground, ranging from the banks of Town Creek and the

Guadalupe to the streets of Kerrville, where the plants were large and branched above.

May 16 (1754); type locality, Bejar, Mexico.

AMBLYOLEPIS DC. Prodr. 5: 667 (1836).
Amblyolepis setigera DC. Prodr. 5: 667 (1836).

Found only near the edge of the bluff on the left bank of the Guadalupe, at Kerrville, altitude 1650 feet. Leaves thick and fleshy, emitting a rather pleasant odor when dried.

May 14 (1746); type locality, Texas.

GAILLARDIA Foug. Mem. Acad. Paris, 1786, 5, *t. 1, 2* (1786).
Gaillardia pulchella Foug. Mem. Acad. Paris, 1786, 5 (1786).

Two forms were collected at Corpus Christi, in low, dry ground along the beach. One is ascending, rather naked, at least below, inclined to be woody, and leaflets usually entire. The other is low, prostrate, with thick, fleshy leaves, and large flowers, with red-brown rays (1424). At San Antonio, and about Kerrville, where it grows in large patches, the ordinary form is found, with the outer part of the range yellow.

April 19 (1584).

Gaillardia suavis (Engelm. & Gray) Britt. & Rusby, Trans. N. Y. Acad. Sci. **7:** 11 (1887).

Agassizia suavis Engelm. & Gray, Proc. Am. Acad. **1:** 49 (1846).
Gaillardia simplex Scheele, Linnaea, **22:** 160 (1849).

At Kerrville, altitude 1700 feet, a few plants were found with ray flowers, which were small, copper-colored. The leaves had a tendency to be more entire than the rayless form, and were somewhat shorter and broader. The ordinary form was quite common.

April 30 (1680); type locality, New Braunfels, Texas.

PTILEPIDA Raf. Am. Month. Mag. **2:** 268 (Feb. 1818).

[ACTINELLA Nutt. Gen. **2:** 173 (1818), not Pers. and not ACTINEA Juss.]

Ptilepida linearifolia (Hook.) Britton, Mem. Torr. Club, **5:** 340 (1894).

Hymenoxys (?) *linearifolia* Hook. Icon. Pl. *t. 146* (1837).
Actinella linearifolia T. & G. Fl. N. A. **2:** 383 (1842).

In rich and often shaded ground about Kerrville, growing in patches, altitude 1620-1650 feet.

April 19 (1619); type locality, Texas.

Ptilepida scaposa (DC.) Britton, Mem. Torr. Club, 5: 340 (1894).
Cephalophora scaposa DC. Prodr. 5: 663 (1836).
Actinella scaposa Nutt. Trans. Am. Phil. Soc. (II.) 7: 379 (1841).
Common in dry, stony ground at Kerrville, usually growing in scattered clumps, altitude 1620-1800 feet.
April 19 (1614); type locality, Texas.

SENECIO L. Sp. Pl. 866 (1753).

Senecio lobatus Pers. Syn. 2: 436 (1807).
Senecio lyratus Michx. Fl. Bor. Am. 2: 120 (1803), not L.
Growing in a grassy meadow at the Oso, and also plentiful on the plateau near Corpus Christi.
March 21 (1476); type locality, Carolina.

CENTAUREA L. Sp. Pl. 909 (1753).

Centaurea Americana Nutt. Jour. Acad. Phila. 2: 117 (1821),
Very plentiful in rich, stony ground along the summits of ridges on the plateau, five miles north of Kerrville, altitude 1900 feet. Noticed also along the railroad between Kerrville and San Antonio, and at San Antonio.
May 21 (1774); type locality, Arkansas.

CICHORIACEAE.

PEREZIA Lag. Amoen. Nat. 1: 31 (1811).

Perezia runcinata Lag.; ex Don, Trans. Linn. Soc. 16: 207 (1830).
Clarionea runcinata Don, Trans. Linn. Soc. 16: 207 (1830).
Near Corpus Christi on the plateau, growing under chapparral. Plentiful only at one place along the Texas and Mexican Railway, just outside of the town.
April 9 (1537); type locality, Texas.

ADOPOGON Neck. Elem. 1: 55 (1790).

[KRIGIA Schreb. Gen. Pl. 532 (1791).]

Adopogon occidentalis mutica (T. & G.).
Krigia occidentalis mutica T. & G. Fl. N. A. 2: 468 (1842).
On the plateau in northeastern Kerr county, near Bear Creek, altitude 1900 feet. Common but scattered, growing in post-oak pasture land.
April 30 (1678); type locality, Arkansas.

LEONTODON L. Sp. Pl. 798 (1753).

Leontodon hispidus L. Sp. Pl. 799 (1753).

A plant referred to this species is common at Corpus Christi, growing in the shell deposit. The leaves are flat on the ground and rather thick. Flowers expand only during the forenoon, and do not remain open very long.

March 7 (1401); type locality, Europe.

PINAROPAPPUS Less. Syn. Comp. 143 (1832).

Pinaropappus roseus Less. Syn. Comp. 143 (1832).

Troximon Roemerianum Scheele, Linnaea, 22: 165 (1849).

In stony ground about Kerrville, especially on slopes and banks, altitude 1620–1800 feet. Flowers pale rose color, or purplish.

April 19 (1602); type locality, Texas.

LYGODESMIA Don, Edinb. Phil. Jour. 6: 305 (1829).

Lygodesmia aphylla Texana T. & G. Fl. N. A. 2: 485 (1842).

In open, usually grassy ground, at Kerrville, altitude 1650–1800 feet. The stems usually break off before the deep-seated, large, fleshy root is reached, in digging into the hard, stony ground.

May 14 (1734); type locality, Texas.

SITILIAS Raf. New Fl. N. A. Part 4, 85 (1836).

[PYRRHOPAPPUS DC. Prodr. 7: 144 (1838).]

Sitilias multicaulis (DC.) Greene, Pitt. 2: 179 (1891).

Pyrrhopappus multicaulis DC. Prodr. 7: 144 (1838).

Common along Town Creek and the Guadalupe, in wet or damp ground. Large plants usually weak and slender.

April 28 (1676); type locality, Texas.

Sitilias grandiflora (Nutt.) Greene, Pitt. 2: 180 (1891).

Parkhausia grandiflora Nutt. Jour. Acad. Phila. 7: 69 (1834).

Pyrrhopappus scaposus DC. Prodr. 7: 144 (1838).

At Corpus Christi, in low, grassy land, along the beach, in the southeastern part of the town, where it is plentiful.

March 6 (1387); type locality, Arkansas.

INDEX.

Abutilon	64
Acacia	6, 7, 42, 43
Acacia	44, 45
Acalypha	59
Acerates	78, 79
Achryanthes	32
Actinella	109, 110
Actinomeris	106
Acuan	44, 45
Adhatoda	95
Adiantum	9
Adopogon	110
Æcidium	81
Aizoon	35
Agassizia	109
Agave	25
Agrostis	15
Allionia	34
Allium	23
Allium	23
Alternanthera	32
Amaranthus	31, 32
Amblyolepis	109
Ambrosia	103
Amellus	101
Ammannia	69
Ammoselinum	72
Amorpha	48
Ampelanus	79
Ampelopsis	63, 64
Anantherix	78
Andropogon	10
Antirrhinum	91
Antirrhinum	91
Aphanostephus	101
Aphora	60
Apium	73
Apium	72, 73
Aplopappus	100, 101
Arabis	41
Arenaria	35
Arenaria	36
Argemone	38
Argyrochaeta	103
Argythamnia	60
Aristida	14
Asclepias	77
Asclepias	77, 78
Asclepiodora	78
Aspidium	9
Astragalus	51, 52
Atrema	72
Atriplex	31
Baptisia	46
Barrattia	105
Bartonia	68
Batis	4, 33
Berberis	37, 38
Berlandiera	103
Bifora	7, 71
Bigelowia	101
Boehmeria	29
Boerhavia	34
Bolivaria	76
Borrichia	105
Bouteloua	15, 16
Bowlesia	74
Brauneria	104
Brazoria	7, 88
Briza	17
Bromus	18
Buchloë	16
Bulbilis	16
Bumelia	75
Buphthalmum	105
Caesalpinia	46
Calceolaria	67
Callicarpa	85
Calliopsis	107
Callirhoë	64, 65
Calophanes	94
Calydorea	25
Calylophus	70
Calymenia	34
Calyptocarpus	106
Campanula	99
Canna	26
Cantua	81
Capnoides	38
Capraria	93
Cardamine	39, 40
Cardiospermum	62
Carum	73
Carya	26
Cassia	7, 46
Castela	6, 56
Castilleja	94
Ceanothus	63
Cebatha	38
Celtis	6, 27, 28, 29
Cenchrus	14
Cenchrus	10
Centaurea	110
Cephalantbus	96
Cephalophora	110
Cerasus	42

Ceratachloa	18	Discopleura	73
Cercis	46	Dolichos	53
Cestrum	90	Draba	40, 41
Chaerophyllum	72	Dracocephalum	88
Chamaeraphis	13	Dryopteris	9
Chamaesaracha	89	Dysodia	108
Chenopodium	31	Echinacea	104
Chloris	15	Echinocactus	68
Chondrophora	101	Echinospermum	83
Chondrosium	16	Ehretia	83
Chrysopsis	100	Eleocharis	19
Chthamalia	79	Elusine	16
Chrysopogon	10	Elymus	19
Cienfugosia	67	Encelia	105
Cissus	64	Engelmannia	103
Cladium	20	Enslenia	79
Cladothrix	32	Eragrostis	17, 18
Clarionea	110	Erigeron	102
Clematis	36, 37	Erinus	93
Cocculus	38	Eriocarpum	101
Coelostylis	76	Eriogonum	7, 29
Colubrina	6, 63	Erodium	53
Commelina	21	Erysimum	39
Commelina	22	Erythraea	76, 77
Conobea	93	Eupatorium	100
Convolvulus	80	Euphorbia	60, 61
Cooperia	24, 25	Eustoma	77
Coreopsis	6, 106, 107	Eutoca	82
Coreopsis	107	Evax	102
Cornus	74	Evolvulus	80
Corydalis	38	Eysenhardtia	48
Cosmidium	107	Fagara	56
Crataegus	42	Fedia	98
Cressa	80	Festuca	18
Croton	58	Festuca	18
Crusea	97	Filaginopsis	102
Cucumis	98	Filago	102
Cucurbita	98	Florestina	108
Cupressus	9	Fraxinus	76
Cuscuta	81	Fugosia	67
Cynoglossum	83	Fuirena	20
Cynosciadum	72	Gaillardia	109
Cynosurus	16	Galium	97
Cyperus	19	Galphimia	54
Dactyloctenium	16	Gaura	71
Dalea	49	Gelasine	25
Darlingtonia	44	Gentiana	77
Darwinia	51	Geranium	53
Dasylirion	24	Geranium	53
Daucosma	73	Gilia	81, 82
Daucus	17	Gnaphalium	102
Daucus	73	Gomphrena	32, 58
Delphinium	36	Gomphrena	33
Desmanthus	44, 45	Gonolobus	79
Desmodium	52	Gratiola	93
Dichromena	20	Greenia	15
Dianthera	95	Grindelia	100
Digitaria	12	Guiacum	55
Diodia	97	Hedeoma	85, 86
Diospyros	75	Hedysarum	52
Dipteracanthus	94	Helenium	108

Helianthus	105	Malvastrum	65
Heliosciadum	73	Malvaviscus	67
Heliotropium	83	*Malveopsis*	65
Hendecandra	58	Marilaunidium	82
Herpestis	93	Marshallia	107
Hicoria	26	*Maurandia*	91
Hicorius	26	Maximowicza	99
Hofmanseggia	46	Medicago	47
Holcus	10	Megapterium	70
Hordeum	19	Meibomia	52
Houstonia	96	Melampodium	102
Hydrocotyle	74	Melia	57
Hymenatherum	108	Melica	18
Hymenopappus	107	Melilotus	47
Hymenoxys	109	Menispermum	38
Ilex	62	Menodora	76
Illecebrium	33	Mentzelia	68
Illecebrum	32	Meriolix	70
Indigofera	51	Metastelma	79
Ionidium	67	Mimosa	43
Ipomoea	80	*Mimosa*	42, 44
Ipomoea	91	Mirabilis	34
Iresine	33	Mollugo	35
Jatropha	59	Monarda	8, 87
Juglans	7, 26	*Monechma*	95
Juglans	26	Monniera	93
Juncus	22	Morongia	44
Kallstroemia	35	Morus	28
Krameria	57	Mozinna	59
Krigia	110	*Nama*	82
Kuhnistera	7, 49, 50	*Nasturtium*	39, 40, 41
Lantana	85	Nazia	10
Lappago	10	*Neckeria*	38
Lappula	83	Neptunia	45
Legouzia	99	Nicotiana	90
Leontodon	111	*Nicotiana*	91
Lepidium	39, 81	Nothoscordum	23
Lepachys	104	*Nuttallia*	65
Leptocaulis	73	Œnothera	70
Leptochloa	16	*Œnothera*	70
Lespedeza	52	Oldenlandia	96
Lesquerella	6, 7, 40, 41	Oligogyne	106
Leucopsidium	101	Onosmodium	83, 84
Limnodea	14	Opuntia	68, 69
Limosella	93	Oxalis	54
Linaria	91	*Oxybaphus*	34
Lindheimera	103	*Palafoxia*	108
Linosyris	101	Panicum	11, 12, 13
Linum	54	*Panicum*	13
Lipochaeta	105	Parietaria	29
Lippia	85	*Parkhausia*	111
Lisianthus	77	Parkinsonia	45
Lithospermum	83, 84	Paronychia	36
Lonicera	98	Parosela	49
Lonicera	98	Parthenium	103
Lupinus	46	Paspalum	10, 11
Lycium	81, 89	*Paspalum*	12
Lygodesmia	111	*Pavonia*	67
Lythrum	69	*Pentagonia*	99
Malpighia	55	Pentstemon	91, 92, 93
Malva	65	Perezia	110

Petalostemon	49, 50, 51	Schrankia	44
Petunia	91	Scirpus	19, 20
Phacelia	82	Scutellaria	87, 88
Phalaris	14	Senecio	110
Philoxerus	33	Sesbania	51
Phlox	81	Sesleria	16
Phoradendron	29	Sesuvium	35
Phyllanthus	57	Setaria	13, 14
Physalis	89	Sicydium	99
Physostegia	68	Sida	65, 66
Physostegia	88	*Sida*	64
Phytolacca	33	Sieglingia	16, 17
Pinaropappus	111	Silene	35
Plantago	95, 96	Silphium	103
Platanus	41	Siphonoglossa	95
Pleurobolus	52	Sison	73
Poa	17, 18	Sisymbrium	39
Polanisia	41	*Sisymbrium*	39, 40
Polecarpon	35	Sisyrinchium	25
Polemonium	81	Sitilias	111
Polygala	57	Smilax	24
Polygonum	30, 31	Solanum	90
Polypremum	76	*Solanum*	89
Polypteris	108	Sophora	46
Polytaenia	72	Sorghum	10
Porliera	55	Specularia	99
Portulaca	35	Spergularia	36
Prosopis	5, 6, 29, 45	Spermolepis	73
Prunus	42	Sphaeralcea	67
Psoralea	47, 48	Spigelia	76
Psoralea	49	Stachys	88
Ptelea	56	Stemmodontia	105
Pterota	56	Stevia	108
Ptilepida	109, 110	Stillingia	59
Ptilimnium	73	Suaeda	4, 31
Pyrrhopappus	111	Symphoricarpos	98
Quercus	49	Synedrella	106
Ranunculus	37	*Syntherisma*	12
Rhamnus	63	Synthlipsis	41
Rhamnus	63	Talinum	35
Rhus	61, 62	Taxodium	9
Ridan	106	Tetragonotheca	105
Rivina	33	Teucrium	88, 89
Roripa	39, 41	Thamnosma	55
Rosa	42	Thelesperma	7, 107
Rudbeckia	104	Thryallis	54
Rudbeckia	104	*Thurberia*	15
Ruellia	94	Tillandsia	21
Rumex	30	Tinantia	22
Rutosma	56	Tissa	36
Salix	26	Tradescantia	21
Salvia	86	*Tradescantia*	22
Salviastrum	86	Tragia	59
Samolus	74, 75	*Tragus*	10
Sanicula	72	Tribulus	55
Sapindus	8, 62	*Tricodium*	15
Sarratia	31	Tricuspis	16, 17
Schinus	56	*Trifolium*	47
Schizocarya	71	Triodia	16
Schoenocaulon	24	Trisetum	15
Schoenus	20	Troximon	111

Turritis	41	Vicia	53
Unguadia	62	*Viguiera*	105
Uniola	18	Vincetoxicum	79
Urtica	28	*Viola*	68
Urtica	29	Vitis	63
Usteria	91	*Vitis*	64
Utricularia	94	*Wedelia*	105
Valerianella	98	Xexmenia	105
Verbascum	91	Xylopleurum	70
Verbena	85	Yucca	23, 24
Verbena	84	Zanthoxylum	56
Verbesina	106	*Zimenesia*	106
Vernonia	99, 100	*Zizania*	14
Veronica	64	Zizaniopsis	14
Vesicaria	40	Zizyphus	4
Viburnum	97		

Plate 1.

Rumex spiralis Small.

Plate 2.

Kuhnistera pulcherrima A. A. Heller.

Plate 3.

Samolus alyssoides A. A. Heller.

Asclepias Texana A. A. Heller.

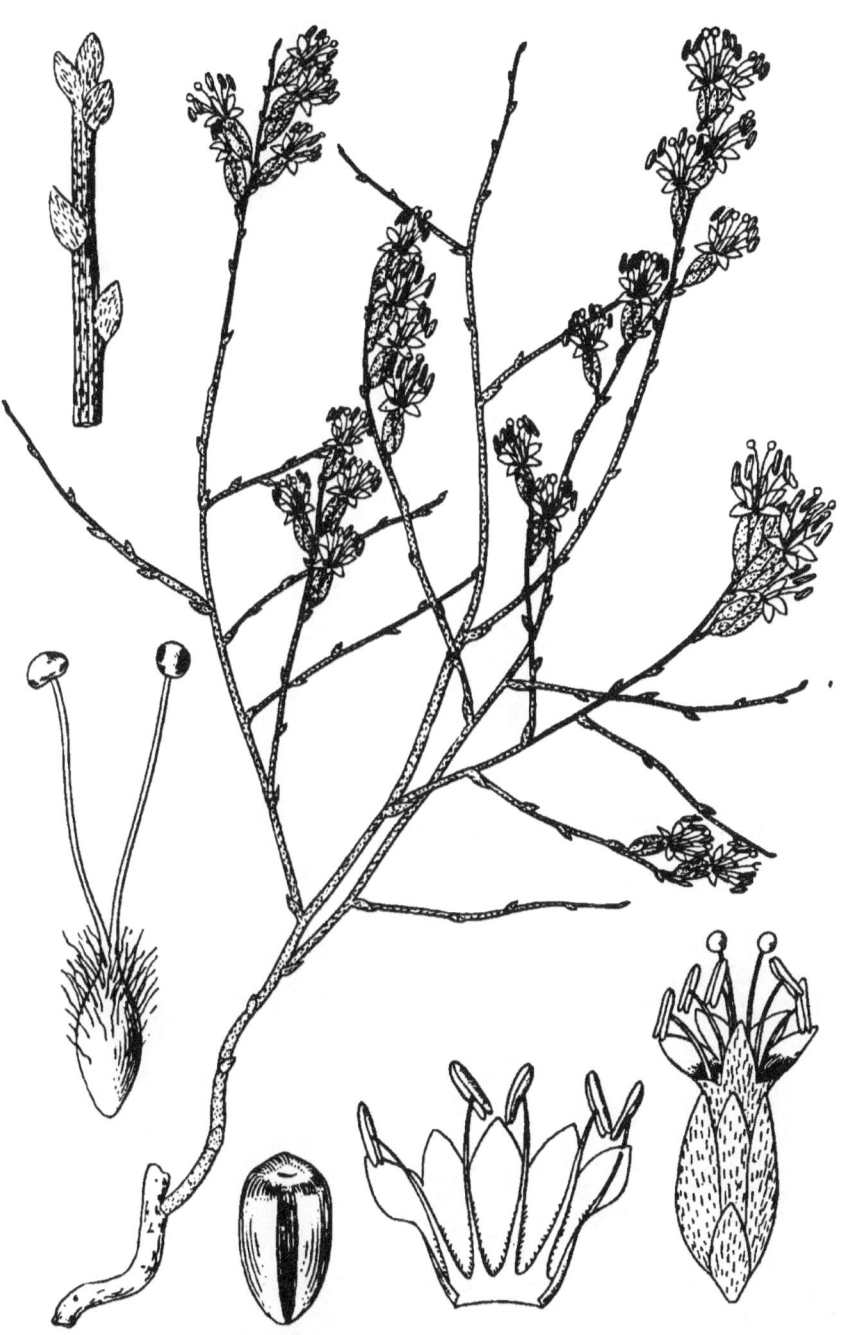

Plate 5.

Cressa aphylla A. A. Heller.

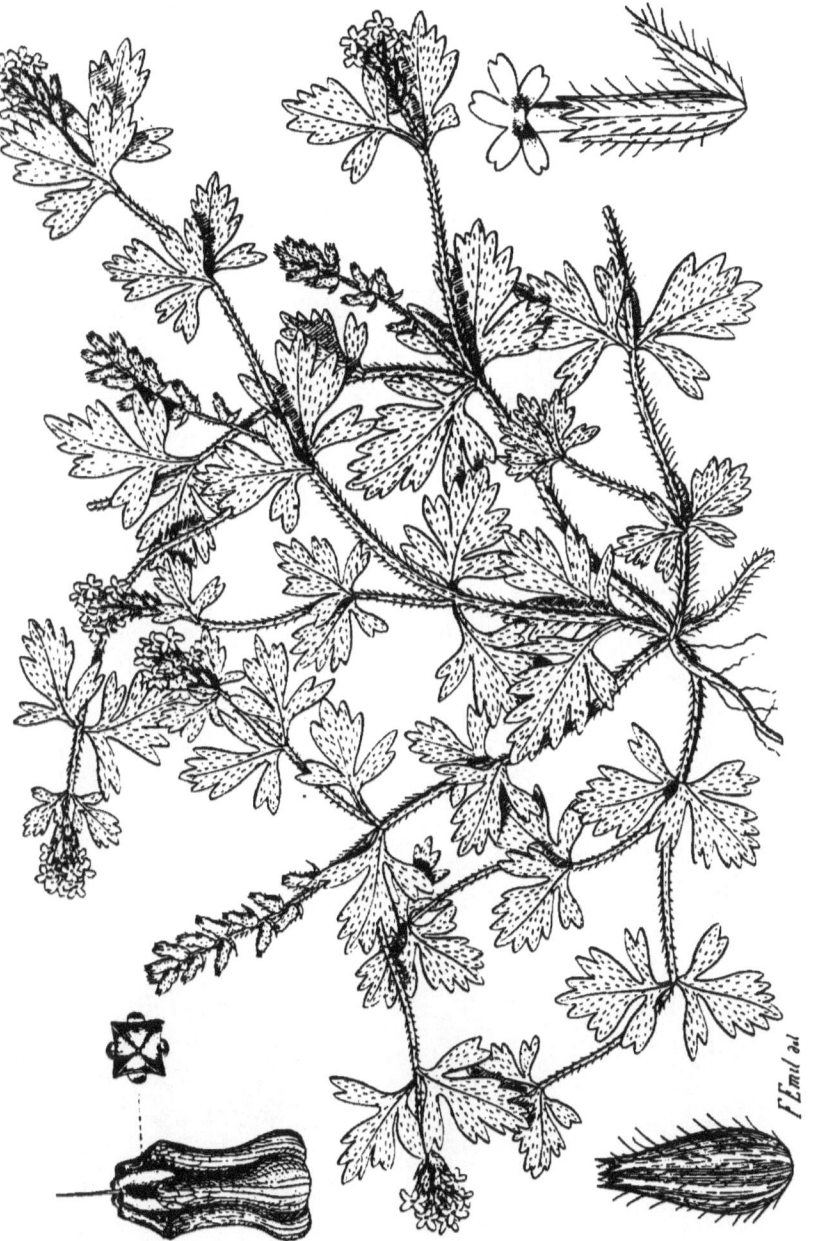

Plate 7.

Penstemon Guadalupensis A. A. Heller.